Introduction to Agricultural Biotechnology

Introduction to Agricultural Biotechnology

Donald Shaffer

MURPHY & MOORE
www.murphy-moorepublishing.com

Introduction to Agricultural Biotechnology
Donald Shaffer
ISBN: 978-1-63987-328-9 (Hardback)

MURPHY & MOORE
Published by Murphy & Moore Publishing,
1 Rockefeller Plaza,
New York City, NY 10020, USA

Cataloging-in-Publication Data

Introduction to agricultural biotechnology / Donald Shaffer.
p. cm.
Includes bibliographical references and index.
ISBN 978-1-63987-328-9
1. Agricultural biotechnology. 2. Biotechnology. 3. Agriculture. I. Shaffer, Donald.
S494.5.B563 I58 2022
338.16--dc23

For more information regarding Murphy & Moore Publishing and its products, please visit the publisher's website www.murphy-moorepublishing.com

Table of Contents

Preface

This book has been written, keeping in view that students want more practical information. Thus, my aim has been to make it as comprehensive as possible for the readers. I would like to extend my thanks to my family and co-workers for their knowledge, support and encouragement all along.

The field of agricultural science which uses different scientific tools and techniques for modifying plants, animals and microorganisms is called agricultural biotechnology. Genetic engineering, molecular diagnostics, vaccines, molecular markers and vaccines are the techniques used in agricultural biotechnology. In crop biotechnology, desired traits are exported from a particular crops species to a different species. Biotechnology in agriculture offers tools for better understanding of crops and to improve their genetic resource management. It studies genes and manipulates their characteristics to increase productivity and achieve better resistance to diseases and insects. This field is used for improving crop's nutritional content. Crop modification techniques used are traditional breeding, polyploidy, mutagenesis, genome editing, protoplast fusion and transgenics. This book elucidates the concepts and innovative models around prospective developments with respect to agricultural biotechnology. It elucidates new techniques and their applications in a multidisciplinary approach. This textbook aims to serve as a resource guide for students and experts alike and contribute to the growth of the discipline.

A brief description of the chapters is provided below for further understanding:

Chapter - What is Agricultural Biotechnology?

Agricultural biotechnology is the branch of agricultural science that uses genetic engineering, molecular diagnostics and markers, tissue culture, etc. for modifying living species for agricultural uses. Some of the crop modification techniques are mutagenesis, polyploidy, transgenics, etc. All the related aspects of agricultural biotechnology have been carefully analyzed in this chapter.

Chapter - Plant Tissue Culture

A combination of techniques that are used for maintaining the growth of plant cells and tissues under sterile conditions in a nutrient rich-medium is called plant tissue culture. It is widely used in plant sciences, forestry and horticulture. This chapter has been carefully written to provide an easy understanding of plant tissue culture.

Chapter - Plant Metabolic Engineering

Plant metabolic engineering redirects the enzymatic reactions for producing new compounds or enhancing the functioning of existing compounds in an organism. Primary and secondary metabolic pathways are studied under its domain. This chapter discusses the concepts associated with plant metabolic engineering in detail.

Chapter – Agriculture Genomics

The usage of genomics and sequencing in agriculture for improving the crop productivity and livestock production is termed as agriculture genomics. Genetic engineering, reverse genetics, somatic fusion, genetically modified crops and plants, etc. are a few of its concepts. The topics elaborated in this chapter will help in gaining a better perspective of agriculture genomics.

Chapter – Biotechnological Approaches for Pest Management

Biotechnology in pest management is the field of science that targets insect-pest control via detailed analysis and engineering of biological processes. Some of the approaches include usage of transgenic plants, gene-encoding toxins, protein inhibitors, etc. This chapter sheds light on various biotechnological approaches for pest management to provide a thorough understanding of the subject.

Chapter – Bioinformatics in Agriculture

Bioinformatics is the discipline of science that deals with the study of information technology to manage biological data for decoding plant genomes. It has aided in genome sequencing for studying protein structures, increasing plant yield, improving nutritional quality, resisting insects and pests, etc. This chapter closely examines the role of bioinformatics in agriculture for an extensive understanding of the subject.

Donald Shaffer

1

What is Agricultural Biotechnology?

Agricultural biotechnology is the branch of agricultural science that uses genetic engineering, molecular diagnostics and markers, tissue culture, etc. for modifying living species for agricultural uses. Some of the crop modification techniques are mutagenesis, polyploidy, transgenics, etc. All the related aspects of agricultural biotechnology have been carefully analyzed in this chapter.

Agricultural biotechnology is the use of different scientific techniques to modify plants and animals. The undesirable characteristics like susceptibility to diseases and low productivity are bred out. If there is a particular trait that the plant or animal can benefit from, it can be bred in by using a gene that contains the characteristic.

Biotechnology has especially been beneficial in improving agricultural productivity and increasing the resistance of plants to diseases. Scientists do this by studying the DNA. They first identify the gene that would be beneficial to the plant or animal then work with the characteristics conferred in a precise and exact manner to achieve the desired outcome.

Advantages and Disadvantages

The use of biotechnology in the field of agriculture does not only allow for crops to grow more and under more difficult circumstances, it can literally make them better. In other words, science allows us to introduce specific genes to increase the nutritional value of crops.

This has been attempted with rice, one of the world's most eaten food, where scientists used genetic engineering to produce rice rich in vitamin A. What scientists noticed is that while rice already contains the genes that produce vitamin A, these get turned off as the rice grow; so what the scientists did was to reverse the process so that the vitamin A genes get activated during the growth.

As such, agritech can help in resolving hunger but also malnutrition. This is therefore not a solution we can afford to ignore when so many hundreds of people, many young children, suffer from malnutrition.

Proponents of agritech, however, believe that their technology can introduce truly sustainable farming practices and even reduce the environmental impact of agriculture.

For example, genetically modified seeds can have improved resistance to germination failure. This allows farmers to plant these seeds without having to till the soil, which is a practice that disturbs beneficial soil organisms, results in the loss of nutrients from soils by bringing them to the surface from where they get washed off by the rain, and releases carbon trapped in the soil organic matter.

What is more, supporters of agritech also believe that their technology can also reduce waste and optimize the food available to us in supermarkets.

This is because genetically modified produce can be given properties that allow it to be harvested when ripe but then the ripening process can be stopped which means consumers can have access to fresher produce with a longer shelf life.

This also minimizes the amount of food that could go back before even reaching retailers.

But if there are so many benefits of agricultural biotechnology, what is holding back further pick-up of these practices by farmers?

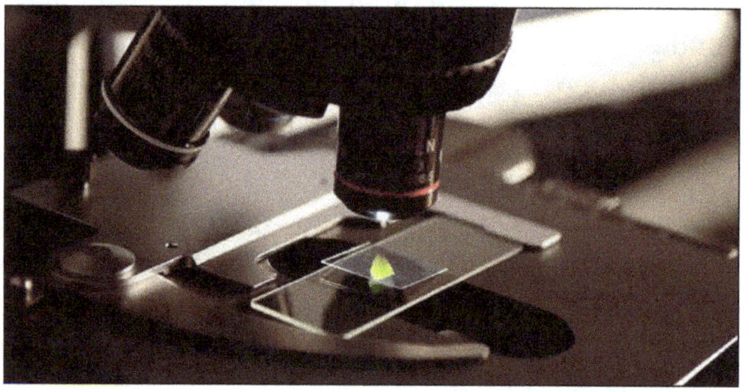

Disadvantages of using Biotechnology in Agriculture

While we have genetically modified our crops and breeds for centuries, there is some hesitation about the more intrusive albeit more targeted way biotechnology does this.

Another aspect that is worth noting is that this genetic engineering is so extreme that we are essentially taking genes from one organism and inserting them into a completely different organism; in all probability, this would not have happened using the traditional approaches used by farmers before.

Perhaps part of the apprehension is owed to the fact that genetically modified food has only been available since the early 1990s.

Given the relatively short time they have been cultivated, scientists cannot draw firm conclusions about their potential long term effects to the environment and health.

However, some scientists have found that genetically modified plants have altered the life span, disease process and cognitive abilities of insects that feed on these plants.

The extensive use of agritech has also created some fear among agriculturalists that it could lead to a decrease in biodiversity. If farmers were to find a particular crop more profitable and easy to grow, farmers would likely switch to this more profitable crop and abandon other varieties.

The modified crop would dominate and other local varieties would be marginalized or even grow extinct. Biodiversity loss is a dangerous trend as it weakens the health of our ecosystems, puts food security at risk and minimizes our ability respond to climate change.

From a human health point of view, scientists are concerned that genetically modified food may create new allergens. This is why the United States Food and Drug Administration regulations for genetically modified food also includes extensive allergenic tests.

Even with all the evidence on the table, it is difficult to take a firm decision about the application of biotechnology in agriculture.

What is certain is that we need to remain vigilant regarding the health and environmental implications. If agritech companies are truly committed to helping people around the globe to escape hunger and support sustainable farming practices, then they should certainly acknowledge the need to ensure that biotechnology products deliver on that without compromising our environment or health.

ETHICS AND AGRICULTURAL BIOTECHNOLOGY

Through the advancement of technology, scientists have been able to develop more precise and powerful tools to produce crops and animals with selected traits that aim to benefit farmers and consumers. Similar to other emerging technologies, biotechnology has instigated worldwide debate and confusion as a result of mixed messages from various people - be they scientists, academics, critics, industry, religious representatives or consumer bodies. The worldwide debates on the pros and cons of biotechnology have been likened to a battleground and a prominent place for virtually every ethical concern. It has stirred conflicting ideas and opinions and has polarized sectors not only among stakeholders but even between countries.

While agriculture has long been a topic of philosophical, religious and political reflection, it is only in the late 20th century that systematic thinking about the values and norms associated with the food system, such as farming, food processing, distribution, trade, and consumption, began to be discussed in the context of agricultural ethics. In addition, by placing biotechnology in the light of globalization, societal debate has moved towards a discussion of ethical and social impacts.

In 2000, the Council of Europe Parliamentary Assembly recommended that it was increasingly important to include ethical considerations centered on humankind, society and the environment in deliberations regarding developments and applications in biotechnology, life sciences, and technology. A year later, the United Kingdom's Royal Society Report asserted that "public debate about genetically modified food must take account of wider issues than the science alone".

What is Agricultural Ethics?

In general, 'ethics' is defined as the ideals, values or standards that people use to determine whether their actions are good or bad. It is what society uses to judge whether an issue or thing is acceptable and justifiable and determines responsibility and justice. Ethics provide guidelines that help one decide what is the right thing to do.

On one hand, ethics is a set of universal norms that are documented through legal or professional codes of practice, religious texts, literature and philosophy. On the other hand, ethics are values defined by a person or groups that are personal, introspective, and hence, difficult to manage for public discussion. Discussion within the agricultural realm is necessary to determine what is right and wrong, what moral standard is or should be used, and why it is the proper one to justify singular or collective acts.

Ethics in agricultural biotechnology therefore encompass value judgments that cover the production, processing, and distribution of food and agricultural products. The Food and Agriculture Organization of the United Nations asserts that ethical values determine its reason for being these being the values for food, enhanced well-being, human health, natural resources, and nature.

Ultimately the goal of agricultural ethics is to "discover or develop clear, noncontradictory, comprehensive, and universal standards for judging right and wrong actions and policies."

What are some Ethical Issues Raised about Agricultural Biotechnology?

Many of the ethical issues that form part of the biotechnology debate can apply also to food and agricultural systems in general. Accepting the need to understand and tolerate societal norms or beliefs, many statements of concern are often general and broad with little explanation about what makes them disagreeable or wrong.

Playing God

Genetic modification is said to involve human intervention into creation and hence, is an unnatural act. Often viewed as a religious question, it avers that the technology is "so intrusive to life processes that they amount to a form of disrespect for humanity's proper relationship to nature, a form of playing God". Some religions ascribe a particular "essence" to each living organism and hence, connect the concept of gene with the idea of essence. Others believe that biotechnology disrupts natural order and violates the limits of what humans are ethically permitted to do. Alternatively, there is the view that science and progress are good things and are God-given faculties to help mankind support life and better manage the environment.

General Welfare and Sustainability

A central issue is whether the technology considers the pursuit of the greatest good together with the concept of sustainability for farmers and the environment. While a technology can provide more food it should not be to the detriment of the environment or to human health or disrupt traditional behavioral systems. In like manner, it is an ethical issue if food that can provide more and better nutrition is not made available to those who need it most. Hence, not to use a technology that has potential to improve the quality of lives of people is also a moral issue. As an environmental issue, questions raised have to do with concerns regarding environmental protection, sustainable use of biodiversity, economic growth and social equity.

Distribution of Benefits and Burdens

A concern particularly in developing countries is the concept of just distribution. Questions have to do with whether the products produced by the technology will be able to provide for those who really need them and whether they will generate wealth for the society as a whole. A technology's ability to increase or decrease the gap between the rich and poor renders it an ethical issue. This includes allegations that products derived from modern biotechnology are being introduced by private companies that have an obligation to make profits. Also up for discussion is whether a technology, while able to increase technical employment might eliminate subsistence labor as a result of replacing cultural operations.

Other concerns include exploitation or control over genetic resources, consumers' choice and rights, and use of genetically modified animals.

Religion and Agricultural Biotechnology

The religious sector, notably the Roman Catholic Church and the Muslim faith, have voiced their views on biotechnology. Islamic scholars note that Islam and science are complementary and Islam supports beneficial scientific innovations to address food security.

Biotechnology, in particular, becomes an issue when it entails a discourse on food. Any GM food must meet the general criterion of halalan tayyiban which means "permissible from the shariah perspective (halal) and of good quality (tayyib)". In Malaysia, there is a fatwa (religious decree) that states that GM foods with DNA from pigs are haram (not permissible) for Muslims to eat. To date, only this fatwa has been issued.

The Jubilee of the Agricultural World Address of John Paul II in 2000 mentioned that in agricultural production or in the case of biotechnology, it must not be evaluated solely on the basis of immediate economic interest but through rigorous scientific and ethical examination. By October 2004, the Pontifical Council for Justice and Peace released the Compendium of the Social Doctrine of the Church which is an "overview of the fundamental framework of the doctrinal corpus of Catholic social teaching." Biotechnology is mentioned as having powerful social, economic, and political impact but that it should be used with prudence, objectivity, and responsibility.

How do we Deal with Ethical Issues?

There is no single set of ethical principles sufficient for building a more equitable and ethical food and agricultural system. However, it recommends the following actions that individuals, states, corporations and voluntary organizations in the international community can take:

- Creating the mechanisms to balance interests and resolve conflicts.
- Supporting and encouraging broad stakeholder participation in policies, programs, and projects.
- Encouraging individuals, communities and nations to engage in dialogue, and ultimately, to do what is ethical.
- Developing and disseminating widely the information and analyses necessary to make wise and ethical decisions.
- Ensuring that decision-making procedures in international food and agriculture policy are well understood and transparent.
- Fostering the use of science and technology in support of a more just and equitable food and agriculture system.
- Ensuring that programs, policies, standards and decisions always take ethical considerations into account so as to lead to enhanced well-being, environmental protection and improved health.
- Developing codes of ethical conduct where they do not currently exist.
- Periodically reviewing ethical commitments and determining whether or not they are appropriate, in the light of new knowledge and changes in circumstances.

CAST suggests the need to institutionalize agricultural ethics. This involves a deliberate move to include some consideration of ethics in the actions, decisions, and policies that stakeholders in the food system create or support. Each stakeholder has to "accept the fact that that if ethical issues are going to be understood, and if ethical conflicts are going to be resolved, it is our responsibility, within the limits of our place in the system, to understand and contribute."

Despite the diversity of ethical issues in agricultural biotechnology, there is a need to understand beliefs and doctrines as this allows coexistence within and across societies, and prevents social conflict. A technology's acceptance is based not only on technological soundness but on how it is perceived to be socially, politically, and economically feasible from the viewpoint of disparate groups. An understanding of ethics helps determine what information is needed by society and how to deal with different opinions. A process of negotiation based on trust is essential to enable stakeholders to participate in debates and decision making.

PRODUCTS OF AGRICULTURAL BIOTECHNOLOGY

When it was first introduced, biotechnology was predominantly used in medicine to research and produce pharmaceutical and diagnostic products that help in preventing and curing diseases. However, over the last few decades, this technology has found a place in the agricultural industry like never before. Years of research indicate that agricultural biotechnology is a safe and beneficial technology that plays a big role in promoting economic and environmental sustainability. Genetically modified crops and food have been the main areas of focus for this technology. However, agro-biotechnology is still producing a variety of other products that offer innumerable benefits to the global population.

Vaccines

New vaccines employing green biotechnology innovations are changing the processes of preventing illnesses, particularly in developing countries. Genetically modified crops have had a significant contribution in the development of vaccines. Foods such as fruits, grains, and vegetables are engineered to carry antigenic proteins which are extracted from pathogens. When injected into the body, these antigens trigger an immune response and boost the resistance of the body against the pathogens.

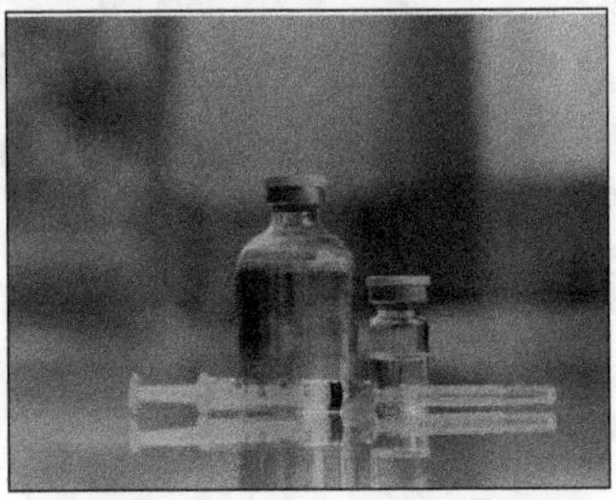

An example is the anti-lymphoma vaccine that's obtained from tobacco. Tobacco plants are engineered to carry RNA from malignant B-cells. The extracted protein is injected into the body, an immune response is triggered which destroys the cancerous cells.

Plant and Animal Reproduction

The use of traditional techniques such as cross-pollination, grafting, and cross-breeding to enhance the behavioral patterns of plants and animals is time-consuming. Agro-biotech has made it possible to enhance plant and animal traits on a molecular level through over-expression or gene removal, or the introduction of foreign genes.

Artificial insemination, embryo transfer, and other associated technologies are used in managing the reproductive functions of an animal and influencing the traits of the resultant offspring. These improvements have increased agricultural productivity in developing countries and enhanced their capabilities to sustain the growing population.

Antibiotics

Agricultural biotechnology is applied in the production of antibiotics for both humans and animals. Animal antibiotics produced through this technology are low cost but equally as efficient as traditionally manufactured antibiotics. Since these antibiotics are obtained from plants, a large quantity of the product can be obtained at a time. Additionally, there is ease of purification and the risk of contamination is minimized as compared to the use of mammalian cells and culture media in antibiotics production.

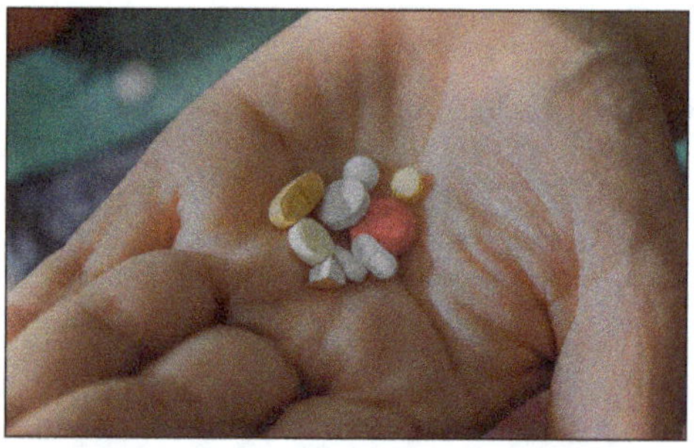

Nutritional Supplements

In a bid to promote better human health globallys, scientists have come up with ways to create genetically modified foods with nutrients that can help fight disease and starvation. A great example of such foods is the golden rice which contains beta carotene, a major source of Vitamin A in the body.

The name of the rice comes from the color of the transgenic grain made from three genes: two from daffodils and one from bacterium. The genes are cloned to make the rice "golden." People who eat this rice supplement their diet with the vitamin and other nutrients that they may not be getting from other foods.

Pesticide-resistant Crops

In the past, farmers have incurred significant losses due to the use of pesticides that affect both crops and weeds. Biotechnology has led to the engineering of plants that are resistant to pesticides. This allows farmers to selectively kill weeds without harming their crop. A famous example is the Roundup-Ready tech introduced by Monsato.

The tech was first introduced in genetically modified soy beans, making them resistant to the herbicide glyphosate. The herbicide can be applied in copious amounts to eliminate other plants on a field other than the Roundup-Ready plants. Selective elimination of weeds saves farmers' valuable time as compared to traditional methods of weeding.

Pest-resistant Crops

For many years, a microbe known as Bacillus thuringiensis (Bt) has been used to dust crops by producing toxic proteins against pests. One of such toxic proteins used for dusting crops is the European corn borer. Scientists have come up with a way to eliminate the use of Bt by introducing pest resistant crops. These are known as Bt crops as the gene that's introduced in the crop was originally identified in Bacillus thuringiensis. Examples if pest resistant crops today are Bt maize, potato, and corn. This toxic protein is only harmful to pests, but is safe for humans. It has saved farmers from dealing with expensive pest infestations in crops.

Flowers

Agricultural biotechnology is not just about developing drugs and genetically modified foods and crops – it has some aesthetic applications as well. Scientists are using gene recognition and transfer techniques to improve the color, size, smell, and other properties of flowers. The technology has also been used to improve other ornamental plants such as shrubs and trees. Some of the techniques applied are similar to those used on crops. For instance, tropical plants' color confrontation can be enhanced to make it possible for the tree to thrive in gardens in the northern regions.

Bio-fuels

The agricultural industry plays a major role in the production of bio-fuels to the extent that feedstock is used for fermentation and purification of bio-oil, bio-ethanol, and bio-diesel. Genetic engineering and enzyme optimization techniques are used to produce good-quality feed stocks for more efficient conversions and higher BTU outputs of the resultant bio-fuels. High-energy and high-yielding feed stocks can reduce the relative costs of harvesting and transportation. The result is high-quality bio-fuel products.

A Biotic Strain Confrontation

A very small proportion of the earth, approximately 20 percent, is arable land. However, scientists have come up with ways to modify crops that can endure conditions such as salinity, cold, and drought. For instance, the detection of genes in plants that are tasked with the uptake of sodium has led to the introduction of plants that can thrive in high-salinity environments.

A technique known as up- or down-regulation of record is used to influence drought-resistance in plants. These technologies have increased food production as plants are able to adapt to hostile climates and non-arable lands.

Manufacture of Power Fibers

The strongest fiber known to man today is spider silk. It is stronger than kevlar, which is used to make bullet-proof vests and has a higher tensile strength than strength. In August 2000, a Canadian company, Nexia, announced that they had engineered transgenic goats that could produce spider silk proteins in their milk.

For a while, the technology seemed like it would solve the problem. However, it was shelved when the scientists couldn't come up with a way to spin the protein into fibers. Although the technology has been put on hold, it is bound to appear again in the future because of high demand for similar products.

2

Plant Tissue Culture

A combination of techniques that are used for maintaining the growth of plant cells and tissues under sterile conditions in a nutrient rich-medium is called plant tissue culture. It is widely used in plant sciences, forestry and horticulture. This chapter has been carefully written to provide an easy understanding of plant tissue culture.

Tissue culture is the in vitro aseptic culture of cells, tissues, organs or whole plant under controlled nutritional and environmental conditions often to produce the clones of plants. The resultant clones are true-to type of the selected genotype. The controlled conditions provide the culture an environment conducive for their growth and multiplication. These conditions include proper supply of nutrients, pH medium, adequate temperature and proper gaseous and liquid environment.

Plant tissue culture technology is being widely used for large scale plant multiplication. Apart from their use as a tool of research, plant tissue culture techniques have in recent years, become of major industrial importance in the area of plant propagation, disease elimination, plant improvement and production of secondary metabolites. Small pieces of tissue (named explants) can be used to produce hundreds and thousands of plants in a continuous process. A single explant can be multiplied into several thousand plants in relatively short time period and space under controlled conditions, irrespective of the season and weather on a year round basis. Endangered, threatened and rare species have successfully been grown and conserved by micro-propagation because of high coefficient of multiplication and small demands on number of initial plants and space.

In addition, plant tissue culture is considered to be the most efficient technology for crop improvement by the production of somaclonal and gametoclonal variants. The micro-propagation technology has a vast potential to produce plants of superior quality, isolation of useful variants in well-adapted high yielding genotypes with better disease resistance and stress tolerance capacities. Certain type of callus cultures give rise to clones that have inheritable characteristics different from those of parent plants due to the possibility of occurrence of somaclonal variability, which leads to the development of commercially important improved varieties. Commercial production of plants through micro-propagation techniques has several advantages over the traditional methods of propagation through seed, cutting, grafting and air-layering etc. It is rapid propagation processes that can lead to the production of plants virus free. Coryodalisyanhusuo, an important medicinal plant was propagated by somatic embryogenesis from tuber-derived callus to produce disease free tubers. Meristem tip culture of banana plants devoid from banana bunchy top virus (BBTV) and brome mosaic virus (BMV) were produced. Higher yields have been obtained by culturing pathogen free germplasmin vitro. Increase in yield up to 150% of virus-free potatoes was obtained in controlled conditions.

Basics of Plant Cell and Tissue Culture

In plant cell culture, plant tissues and organs are grown in vitro on artificial media, under aseptic and controlled environment. The technique depends mainly on the concept of totipotentiality of plant cells which refers to the ability of a single cell to express the full genome by cell division. Along with the totipotent potential of plant cell, the capacity of cells to alter their metabolism, growth and development is also equally important and crucial to regenerate the entire plant. Plant tissue culture medium contains all the nutrients required for the normal growth and development of plants. It is mainly composed of macronutrients, micronutrients, vitamins, other organic components, plant growth regulators, carbon source and some gelling agents in case of solid medium. Murashige and Skoog medium (MS medium) is most extensively used for the vegetative propagation of many plant species in vitro. The pH of the media is also important that affects both the growth of plants and activity of plant growth regulators. It is adjusted to the value between 5.4 - 5.8. Both the solid and liquid medium can be used for culturing. The composition of the medium, particularly the plant hormones and the nitrogen source has profound effects on the response of the initial explant.

Plant growth regulators (PGR's) play an essential role in determining the development pathway of plant cells and tissues in culture medium. The auxins, cytokinins and gibberellins are most commonly used plant growth regulators. The type and the concentration of hormones used depend mainly on the species of the plant, the tissue or organ cultured and the objective of the experiment. Auxins and cytokinins are most widely used plant growth regulators in plant tissue culture and their amount determined the type of culture established or regenerated. The high concentration of auxins generally favors root formation, whereas the high concentration of cytokinins promotes shoot regeneration. A balance of both auxin and cytokinin leads to the development of mass of undifferentiated cells known as callus.

Maximum root induction and proliferation was found in Stevia rebaudiana, when the medium is supplemented with 0.5 mg/l NAA. Cytokinins generally promote cell division and induce shoot formation and axillary shoot proliferation. High cytokinin to auxin ratio promotes shoot proliferation while high auxin to cytokinins ratio results in root formation. Shoot initiation and proliferation was found maximum, when the callus of black pepper was shifted to medium supplemented with BA at the concentration of 0.5 mg/l. Gibberellins are used for enhanced growth and to promote cell elongation.

Maximum shoot length was observed in Phalaenopsis orchids when cultured in medium containing 0.5 mg/l GA_3.

Tissue Culture in Agriculture

As an emerging technology, the plant tissue culture has a great impact on both agriculture and industry, through providing plants needed to meet the ever increasing world demand. It has made significant contributions to the advancement of agricultural sciences in recent times and today they constitute an indispensable tool in modern agriculture.

Biotechnology has been introduced into agricultural practice at a rate without precedent. Tissue culture allows the production and propagation of genetically homogeneous, disease-free plant material. Cell and tissue in vitro culture is a useful tool for the induction of somaclonal variation. Genetic variability induced by tissue culture could be used as a source of variability to obtain new stable genotypes. Interventions of biotechnological approaches for in vitro regeneration, mass micro-propagation techniques and gene transfer studies in tree species have been encouraging. In vitro cultures of mature and/or immature zygotic embryos are applied to recover plants obtained from inter-generic crosses that do not produce fertile seeds. Genetic engineering can make possible a number of improved crop varieties with high yield potential and resistance against pests. Genetic transformation technology relies on the technical aspects of plant tissue culture and molecular biology for:

- Production of improved crop varieties.
- Production of disease-free plants (virus).
- Genetic transformation.
- Production of secondary metabolites.
- Production of varieties tolerant to salinity, drought and heat stresses.

Germplasm Conservation

In vitro cell and organ culture offers an alternative source for the conservation of endangered genotypes. Germplasm conservation worldwide is increasingly becoming an essential activity due to the high rate of disappearance of plant species and the increased need for safeguarding the floristic patrimony of the countries. Tissue culture protocols can be used for preservation of vegetative tissues when the targets for conservation are clones instead of seeds, to keep the genetic background of a crop and to avoid the loss of the conserved patrimony due to natural disasters, whether biotic or abiotic stress. The plant species which do not produce seeds (sterile plants) or which have 'recalcitrant' seeds that cannot be stored for long period of time can successfully be preserved via in vitro techniques for the maintenance of gene banks.

Cryopreservation plays a vital role in the long-term in vitro conservation of essential biological material and genetic resources. It involves the storage of in vitro cells or tissues in liquid nitrogen that results incryo-injury on the exposure of tissues tophysical and-chemical stresses. Successful cryopreservation is often ascertained by cell and tissue survival and the ability to re-grow or regenerate into complete plants or form new colonies. It is desirable to assess the genetic integrity of recovered germplasm to determine whether it is 'true-to-type' following cryopreservation. The fidelity of recovered plants can be assessed at phenotypic, histological, cytological, biochemical and molecular levels, although, there are advantages and limitations of the various approaches used to assess genetic stability. Cryobionomics is a new approach to study genetic stability in the cryopreserved plant materials. The embryonic tissues can be cryopreserved for future use or for germplasm conservation.

Embryo Culture

Embryo culture is a type of plant tissue culture that is used to grow embryos from seeds and ovules in a nutrient medium. In embryo culture, the plant develops directly from the embryo or indirectly through the formation of callus and then subsequent formation of shoots and roots. The technique has been developed to break seed dormancy, test the vitality of seeds, production of rare species and haploid plants. It is an effective technique that is employed to shorten the breeding cycle of plants by growing excised embryos and results in the reduction of long dormancy period of seeds. Intra-varietal hybrids of an economically important energy plant "Jatropha" have been produced successfully with the specific objective of mass multiplication. Somatic embryogenesis and plant regeneration has been carried out in embryo cultures of Jucara Palm for rapid cloning and improvement of selected individuals. In addition, conservation of endangered species can also be attained by practicing embryo culture technique. Recently a successful protocol has been developed for the in vitro propagation of Khayagrandifoliola by excising embryos from mature seeds. The plant has a high economic value for timber wood and for medicinal purposes as well. This technique has an important application in forestry by offering a mean of propagation of elite individuals where the selection and improvement of natural population is difficult.

Genetic Transformation

Genetic transformation is the most recent aspect of plant cell and tissue culture that provides the mean of transfer of genes with desirable trait into host plants and recovery of transgenic plants. The technique has a great potential of genetic improvement of various crop plants by integrating in plant biotechnology and breeding programmes. It has a promising role for the introduction of agronomically important traits such as increased yield, better quality and enhanced resistance to pests and diseases.

Genetic transformation in plants can be achieved by either vector-mediated (indirect gene transfer) or vectorless (direct gene transfer) method. Among vector dependant gene transfer methods, Agrobacterium-mediated genetic transformation is most widely used for the expression of foreign genes in plant cells. Successful introduction of agronomic traits in plants was achieved by using root explants for the genetic transformation. Virus-based vectors offers an alternative way of stable and rapid transient protein expression in plant cells thus providing an efficient mean of recombinant protein production on large scale.

Recently successful transgenic plants of Jatropha were obtained by direct DNA delivery to mature seed-derived shoot apices via particle bombardment method. This technology has an important impact on the reduction of toxic substances in seeds thus overcoming the obstacle of seed utilization in various industrial sector. Regeneration of disease or viral resistant plants is now achieved by employing genetic transformation technique. Researchers succeeded in developing transgenic plants of potato resistant to potato virus Y (PVY) which is a major threat to potato crop worldwide. In addition, marker free transgenic plants of Petunia hybrida were produced using multi-auto-transformation (MAT) vector system. The plants exhibited high level of resistance to Botrytis cinerea,- causal agent of gray mold.

Protoplast Fusion

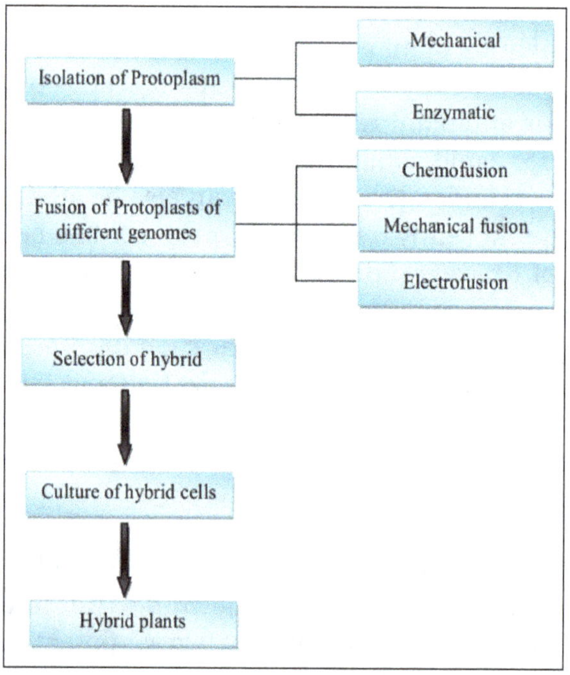

Production of hybrid plant via protoplast fusion.

Somatic hybridization is an important tool of plant breeding and crop improvement by the production of interspecific and intergeneric hybrids. The technique involves the fusion of protoplasts of two different genomes followed by the selection of desired somatic hybrid cells and regeneration of hybrid plants. Protoplast fusion provides an efficient mean of gene transfer with desired trait from one species to another and has an increasing impact on crop improvement. Somatic hybrids were produced by fusion of protoplasts from rice and ditch reed using electrofusion treatment for salt tolerance.

In vitro fusion of protoplast opens a way of developing unique hybrid plants by overcoming the barriers of sexual incompatibility. The technique has been applicable in horticultural industry to create new hybrids with increased fruit yield and better resistance to diseases. Successful viable hybrid plants were obtained when protoplasts from citrus were fused with other related citrinae species. The potential of somatic hybridization in important crop plants is best illustrated by the production of intergeneric hybrid plants among the members of Brassicaceae. To resolvethe problem of loss of chromosomes and decreased regeneration capacity, successful protocol has been established for the production of somatic hybrid plants by using two types of wheat protoplast as recipient and protoplast of Haynaldiavillosa as a fusion donor. It is also employed as an important gene source for wheat improvement.

Haploid Production

The tissue culture techniques enable to produce homozygous plants in relatively short time period through the protoplast, anther and microspore cultures instead of conventional breeding.

Haploids are sterile plants having single set of chromosomes which are converted into homozygous diploids by spontaneous or induced chromosome doubling. The doubling of chromosomes restores the fertility of plants resulting in production of double haploids with potential to become pure breeding new cultivars. The term androgenesis refers to the production of haploid plants from young pollen cells without undergoing fertilization. Sudherson et al. Reported haploid plant production of sturt's desert pea by using pollen grains as primary explants. The haploidy technology has now become an integral part of plant breeding programs by speeding up the production of inbred lines and overcoming the constraints of seed dormancy and embryo non-viability. The technique has a remarkable use in genetic transformation by the production of haploid plants with induced resistance to various biotic and abiotic stresses. Introduction of genes with desired trait at haploid state followed by chromosome doubling led to the production of double haploids inbred wheat and drought tolerant plantswere attained successfully.

Current and Future Status of Plant Tissue Culture

The past decades of plant cell biotechnology has evolved as a new era in the field of biotechnology, focusing on the production of a large number of secondary plant products. During the second half of the last century the development of genetic engineering and molecular biology techniques allowed the appearance of improved and new agricultural products which have occupied an increasing demand in the productive systems of several countries worldwide. Nevertheless, these would have been impossible without the development of tissue culture techniques, which provided the tools for the introduction of genetic information into plant cells. Nowadays, one of the most promising methods of producing proteins and other medicinal substances, such as antibodies and vaccines, is the use of transgenic plants. Transgenic plants represent an economical alternative to fermentation-based production systems. Plant-made vaccines or antibodies (plantibodies) are especially striking, as plants are free of human diseases, thus reducing screening costs for viruses and bacterial toxins. The number of farmers who have incorporated transgenic plants into their production systems in 2008 was 13.3 million, in comparison to 11 million in 2007.

Techniques of Plant Tissue Culture

Micro-propagation

Micro-propagation starts with the selection of plant tissues (explant) from a healthy, vigorous mother plant. Any part of the plant (leaf, apical meristem, bud and root) can be used as explant. The whole process can be summarized into the following stages as shown in figure.

Stage 0: Preparation of Donor Plant

Any plant tissue can be introduced in vitro. To enhance the probability of success, the mother plant should be ex vitro cultivated under optimal conditions to minimize contamination in the in vitro culture.

Stage I: Initiation Stage

In this stage an explant is surface sterilized and transferred into nutrient medium. Generally, the combined application of bactericide and fungicide products is suggested. The selection of products depends on the type of explant to be introduced. The surface sterilization of explant in chemical solutions is an important step to remove contaminants with minimal damage to plant cells. The most commonly used disinfectants are sodium hypochlorite, calcium hypochlorite, ethanol and mercuric chloride ($HgCl_2$). The cultures are incubated in growth chamber either under light or dark conditions according to the method of propagation.

Stage II: Multiplication Stage

The aim of this phase is to increase the number of propagules. The number of propagules is multiplied by repeated subcultures until the desired (or planned) number of plants is attained.

Stage III: Rooting Stage

The rooting stage may occur simultaneously in the same culture media used for multiplication of the explants. However, in some cases it is necessary to change media, including nutritional modification and growth regulator composition to induce rooting and the development of strong root growth.

Stage IV: Acclimatization Stage

At this stage, the in vitro plants are weaned and hardened. Hardening is done gradually from high to low humidity and from low light intensity to high light intensity. The plants are then transferred to an appropriate substrate (sand, peat, compost etc.) and gradually hardened under greenhouse.

Somatic Embryogenesis and Organogenesis

Somatic embryogenesis: is an in vitro method of plant regeneration widely used as an important biotechnological tool for sustained clonal propagation. It is a process by which somatic cells or tissues develop into differentiated embryos. These somatic embryos can develop into whole plants without undergoing the process of sexual fertilization as done by zygotic embryos. The somatic embryogenesis can be initiated directly from the explants or indirectly by the establishment of mass of unorganized cells named callus.

Plant regeneration via somatic embryogenesis occurs by the induction of embryogenic cultures from zygotic seed, leaf or stem segment and further multiplication of embryos. Mature embryos are then cultured for germination and plantlet development, and finally transferred to soil.

Somatic embryogenesis has been reported in many plants including trees and ornamental plants of different families. The phenomenon has been observed in some cactus species. There are various factors that affect the induction and development of somatic embryos in cultured cells. A highly efficient protocol has been reported for somatic embryogenesis on grapevine that showed higher plant regeneration sufficiently when the tissues were cultured in liquid medium. Plant growth regulators play an important role in the regeneration and proliferation of somatic embryos. Highest efficiency of embryonic callus was induced by culturing nodal stem segments of rose hybrids on medium supplemented with various PGR's alone or in combination. This embryonic callus showed high germination rate of somatic embryos when grown on abscisic acid (ABA) alone. Somatic embryogenesis is not only a process of regenerating the plants for mass propagation but also regarded as a valuable tool for genetic manipulation. The process can also be used to develop the plants that are resistant to various kinds of stresses and to introduce the genes by genetic transformation. A successful protocol has been developed for regeneration of cotton cultivars with resistance to Fusarium and Verticillium wilts.

Organogenesis refers to the production of plant organs i.e. roots, shoots and leaves that may arise directly from the meristem or indirectly from the undifferentiated cell masses (callus). Plant regeneration via organogenesis involves the callus production and differentiation of adventitious meristems into organs by altering the concentration of plant growth hormones in nutrient medium. Skoog and Muller were the first who demonstrated that high ratio of cytokinin to auxin stimulated the formation of shoots in tobacco callus while high auxin to cytokinin ratio induced root regeneration.

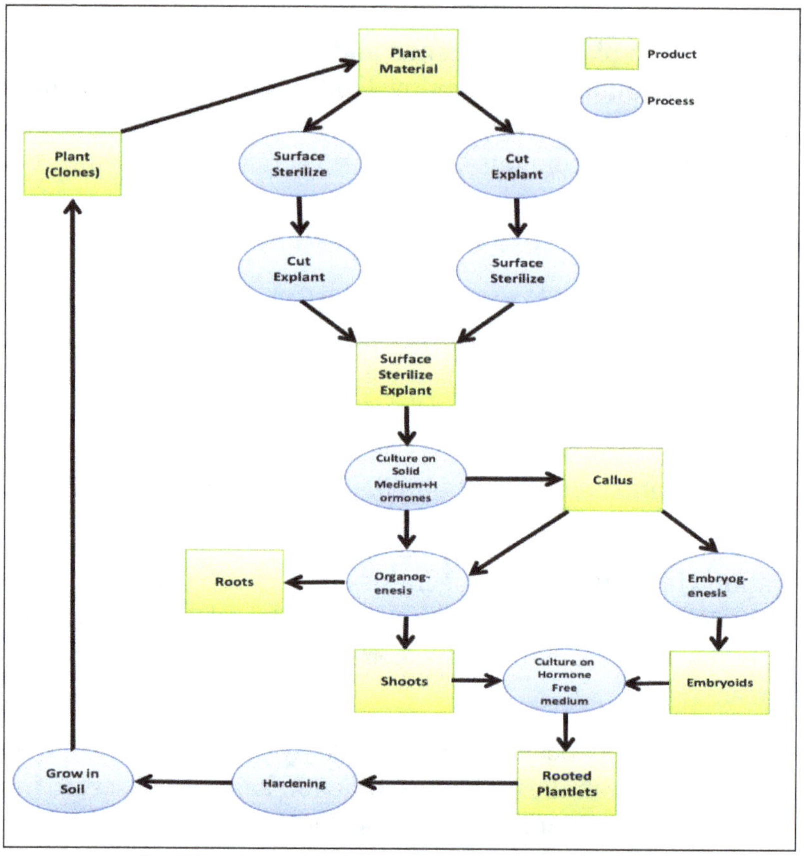

Flow chart summarizing tissue culture experiments.

Tissue Culture in Pharmaceuticals

Plant cell and tissue cultures hold great promise for controlled production of myriad of useful secondary metabolites. Plant cell cultures combine the merits of whole-plant systems with those of microbial and animal cell cultures for the production of valuable therapeutic secondary metabolites. In the search for alternatives to production of medicinal compounds from plants, biotechnological approaches, specifically plant tissue cultures, are found to have potential as a supplement to traditional agriculture in the industrial production of bioactive plant metabolites. Exploration of the biosynthetic capabilities of various cell cultures has been carried out by a group of plant scientists and microbiologists in several countries during the last decade.

Cell suspension culture: Cell suspension culture systems are used now days for large scale culturing of plant cells from which secondary metabolites could be extracted. A suspension culture is developed by transferring the relatively friable portion of the callus into liquid medium and is maintained under suitable conditions of aeration, agitation, light, temperature and other physical parameters. Cell cultures cannot only yield defined standard phytochemicals in large volumes but also eliminate the presence of interfering compounds that occur in the field-grown plants. The advantage of this method is that it can ultimately provide a continuous, reliable source of natural products. The major advantage of the cell cultures include synthesis of bioactive secondary metabolites, running in controlled environment, independently from climate and soil conditions. A number of different types of bioreactors have been used for mass cultivation of plant cells. The first commercial application of large scale cultivation of plant cells was carried out in stirred tank reactors of 200 liter and 750 liter capacities to produce shikonin by cell culture of Lithospermumerythrorhizon. Cell of Catharanthus-roseus, Dioscoreadeltoidea, Digitalis lanata, Panaxnotoginseng, Taxuswallichiana and Podophyllumhexandrum have been cultured in various bioreactors for the production of secondary plant products.

A number of medicinally important alkaloids, anticancer drugs, recombinant proteins and food additives are produced in various cultures of plant cell and tissues. Advances in the area of cell cultures for the production of medicinal compounds has made possible the production of a wide variety of pharmaceuticals like alkaloids, terpenoids, steroids, saponins, phenolics, flavanoids and amino acids. Some of these are now available commercially in the market for example shikonin and paclitaxel (Taxol). Until now 20 different recombinant proteins have been produced in plant cell culture, including antibodies, enzymes, edible vaccines, growth factors and cytokines. Advances in scale-up approaches and immobilization techniques contribute to a considerable increase in the number of applications of plant cell cultures for the production of compounds with a high added value.

Hairy Root Cultures

The hairy root system based on inoculation with Agrobacterium rhizogenes has become popular in the last two decades as a method of producing secondary metabolites synthesized in plant roots. Organized cultures, and especially root cultures, can make a significant contribution in the production of secondary metabolites. Most of the research efforts that use differentiated cultures instead of cell suspension cultures have focused on transformed (hairy) roots. Agrobacterium rhizogenes causes hairy root disease in plants. The neoplastic (cancerous) roots produced by A. rhizogenes infection are characterized by high growth rate, genetic stability and growth in hormone free media. High stability and productivity features allow the exploitation of hairy roots as valuable biotechnological tool for the production of plant secondary metabolites. These genetically transformed root cultures can produce levels of secondary metabolites comparable to that of intact plants. Hairy root technology has been strongly improved by increased knowledge of molecular mechanisms underlying their development. Optimizing the composition of nutrients for hairy root cultures is critical to gain a high production of secondary metabolites.

One of the most important factors governing the growth and morphogenesis of plant tissues in culture is the composition of the culture medium. The basic nutrient requirements of cultured plant cells are very similar to those of whole plants.

Plant tissue and cell culture media are generally made up of some or all of the following components: macronutrients, micronutrients, vitamins, amino acids or other nitrogen supplements, sugar(s), other undefined organic supplements, solidifying agents or support systems, and growth regulators. Several media formulations are commonly used for the majority of all cell and tissue culture work. These media formulations include those described by White, Murashige and Skoog, Gamborg et. al., Schenk and Hilderbrandt, Nitsch and Nitsch, and Lloyd and McCown. Murashige and Skoog's MS medium, Schenk and Hildebrand's SH medium, and Gamborg's B-5 medium are all high in macronutrients, while the other media formulations contain considerably less of the macronutrients.

Macronutrients

The macronutrients provide the six major elements-nitrogen (N), phosphorus (P), potassium (K), calcium (Ca), magnesium (Mg), and sulfur (S)-required for plant cell or tissue growth. The optimum concentration of each nutrient for achieving maximum growth rates varies considerably among species.

Culture media should contain at least 25-60 mM of inorganic nitrogen for adequate plant cell growth. Plant cells may grow on nitrates alone, but considerably better results are obtained when the medium contains both a nitrate and ammonium nitrogen source. Certain species require ammonium or another source of reduced nitrogen for cell growth to occur. Nitrates are usually supplied in the range of 25-20 mm; typical ammonium concentrations range between 2 and 20 mm. However, ammonium concentrations in excess of 8 mm may be deleterious to cell growth of certain species. Cells can grow on a culture medium containing ammonium as the sole nitrogen source if one or more of the TCA cycle acids (e.g., citrate, succinate, or malate) are also included in the culture medium at concentrations of approximately 10 mm. When nitrate and ammonium sources of nitrogen are utilized together in the culture medium, the ammonium ions will be utilized together in the culture medium, the ammonium ions will be utilized more rapidly and before the nitrate ions.

Potassium is required for cell growth of most plant species. Most media contain K, in the nitrate or chloride form, at concentrations of 20-30 mm. The optimum concentrations of P, Mg, S, and Ca range from 1-3 mm when all other requirements for cell growth are satisfied. Higher concentrations of these nutrients may be required if deficiencies in other nutrients exist.

Micronutrients

The essential micronutrients for plant cell and tissue growth include iron (Fe), manganese (Mn), zinc (Zn), boron (B), copper (Cu), and molybdenum (Mo). Chelated forms of iron and zinc are commonly used in preparing culture media. Iron may be the most critical of all the micronutrients. Iron citrate and tartrate may be used in culture media, but these compounds are difficult to dissolve and frequently precipitate after media are prepared. Murashige and Skoog used an ethylene diaminetetraacetic acid (EDTA)-iron chelate to bypass this problem. Cobalt (Co) and iodine (I) may also be added to certain media, but strict cell growth requirements for these elements have not been established. Sodium (Na) and chlorine (Cl) are also used in some media but are not essential for cell growth. Copper and Cobalt are normally added to culture media at concentrations of 0.1 μM, Fe and Mo at 1 μM, I at 5 μM, Zn at 5-30 μM, Mn at 20-90 μM, and B at 25-100 μM.

Carbon and Energy Source

The preferred carbohydrate in plant cell culture media is sucrose. Glucose and fructose may be substituted in some cases, glucose being as effective as sucrose and fructose being somewhat less effective. Other carbohydrates that have been tested include lactose, galactose, rafinose, maltose, and starch. Sucrose concentrations of culture media normally range between 2 and 3 percent. Use of autoclaved fructose can be detrimental to cell growth.

Carbohydrates must be supplied to the culture medium because few plant cell lines have been isolated that are fully autotropic, e.g., capable of supplying their own carbohydrate needs by CO_2 assimilation during photosynthesis.

Vitamins

Normal plants synthesize the vitamins required for their growth and development. Vitamins are required by plants as catalysts in various metabolic processes. When plant cells and tissues are grown in vitro, some vitamins may become limiting factors for cell growth. The vitamins most frequently used in cell and tissue culture media include thiamin (B_1), nicotinic acid, pyridoxine (B_6), and myo-inositol. Thiamin is the one vitamin that is basically required by all cells for growth. Thiamin is normally used at concentrations ranging from 0.1 to 10.0 mg/liter. Nicotinic acid and pyridoxine are often added to culture media but are not essential for cell growth in many species. Nicotinic acid is normally used at concentrations of 0.1-5.0 mg/liter; pyridoxine is used at 0.1-10.0 mg/liter.

Myo-inositol is commonly included in many vitamin stock solutions. Although it is a carbohydrate not a vitamin, it has been shown to stimulate growth in certain cell cultures. Its presence in the culture medium is not essential, but in small quantities myo-inositol stimulates cell growth in most species. Myo-inositol is generally used in plant cell and tissue culture media at concentrations of 50-5000 mg/liter.

Other vitamins such as biotin, folic acid, ascorbic acid, pantothenic acid, vitamin E (tocopherol), riboflavin, and p-aminobenzoic acid have been included in some cell culture media. The requirement for these vitamins by plant cell cultures is generally negligible, and they are not considered growth-limiting factors. These vitamins are generally added to the culture medium only when the concentration of thiamin is below the desired level or when it is desirable to grow cells at very low population densities.

Amino Acids or other Nitrogen Supplements

Although cultured cells are normally capable of synthesizing all of the required amino acids, the addition of certain amino acids or amino acid mixtures may be used to further stimulate cell growth. The use of amino acids is particularly important for establishing cell cultures and protoplast cultures. Amino acids provide plant cells with an immediately available source of nitrogen, which generally can be taken up by the cells more rapidly than inorganic nitrogen. The most common sources of organic nitrogen used in culture media are amino acid mixtures (e.g., casein hydrolysate), L-glutamine, L-asparagine, and adenine. Casein hydrolysate is generally used at concentrations between 0.05 and 0.1 percent. When amino acids are added alone, care must be taken, as they can be inhibitory to cell growth. Examples of amino acids included in culture media to enhance cell growth are glycine at 2 mg/liter, glutamine up to 8 mM, asparagine at 100 mg/liter, L-arginine and cysteine at 10 mg/liter, and L-tyrosine at 100 mg/liter. Tyrosine has been used to stimulate morphogenesis in cell cultures but should only be used in an agar medium. Supplementation of the culture medium with adenine sulfate can stimulate cell growth and greatly enhance shoot formation.

Undefined Organic Supplements

Addition of a wide variety of organic extracts to culture media often results in favorable tissue responses. Supplements that have been tested include protein hydrolysates, coconut milk, yeast extracts, malt extracts, ground banana, orange juice, and tomato juice. However, undefined organic supplements should only be used as a last resort, and only coconut milk and protein hydrolysates are used to any extent today. Protein (casein) hydrolysates are generally added to culture media at a concentration of 0.05-0.1%, while coconut milk is commonly used at 5-20% (v/v).

The addition of activated charcoal (AC) to culture media may have a beneficial effect. The effect of AC is generally attributed to one of three factors: Absorption of inhibitory compounds, absorption of growth regulators from the culture medium, or darkening of the medium. The inhibition of growth in the presence of AC is generally attributed to the absorption of phytohormones to AC. 1-Napthaleneacetic acid (NAA), kinetin, 6-benzylaminopurine (BA), indole-3-acetic acid (IAA), and 6-γ-γ-dimethylallylaminopurine (2iP) all bind to AC, with the latter two growth regulators binding quite rapidly. The stimulation of cell growth by AC is generally attributed to its ability to bind to toxic phenolic compounds produced during culture. Activated charcoal is generally acid-washed prior to addition to the culture medium at a concentration of 0.5-3.0 percent.

Solidifying Agents or Support Systems

Agar is the most commonly used gelling agent for preparing semisolid and solid plant tissue culture media. Agar has several advantages over other gelling agents. First, when agar is mixed with water, it forms a gel that melts at approximately 60°-100 °C and solidifies at approximately 45 °C; thus, agar gels are stable at all feasible incubation temperatures. Additionally, agar gels do not react with media constituents and are not digested by plant enzymes. The firmness of an agar gel is controlled by the concentration and brand of agar used in the culture medium and the pH of the medium. The agar concentrations commonly used in plant cell culture media range between 0.5 and 1.0%; these concentrations give a firm gel at the pH's typical of plant cell culture media.

Another gelling agent commonly used for commercial as well as research purposes is Gelrite. This product is synthetic and should be used at 1.25-2.5 g/liter, resulting in a clear gel which aids in detecting contamination. Alternative methods of support have included use of perforated cellophane, filter paper bridges, filter paper wicks, polyurethane foam, and polyester fleece. Whether explants grow best on agar or on other supporting agents varies from one species of plant to the next.

Growth Regulators

Four broad classes of growth regulators are important in plant tissue culture; the auxins, cytokinins, gibberellins, and abscisic acid. Skoog and Miller were the first to report that the ration of auxin to cytokinin determined the type and extent of organogenesis in plant cell cultures. Both an auxin and cytokinin are usually added to culture media in order to obtain morphogenesis, although the ratio of hormones required for root and shoot induction is not universally the same. Considerable variability exists among genera, species, and even cultivars in the type and amount of auxin and cytokinin required for induction of morphogenesis.

The auxins commonly used in plant tissue culture media are 1H-indole-3-acetic acid (IAA), 1Hindole-3-butyric acid (IBA), (2,4-dichlorophenoxy) acetic acid (2,4-D), and 1-napthaleneacetic acid (NAA). The only naturally occurring auxin found in plant tissues is IAA. Other synthetic auxins that have been used in plant cell culture include 4-chlorophenoxyacetic acid or pchlorophenoxyacetic acid (4 CPA, PCPA), (2,4,5-trichlorophenoxy)acetic acid (2,4,5-T), 3,6- dichloro-2-methoxybenzoic acid (Dicamba), and 4-amino-3,5,6-trichloropicolinic acid (Picloram).

The various auxins differ in their physiological activity and in the extent to which they move through tissue, are bound to the cells, or metabolized. Naturally occurring IAA has been shown to have less physiological activity than synthetic auxins. Based on stem curvature assays, 2, 4-D has eight to twelve times the activity, 2, 4, 5-T has four times the activity, PCPA and Picloram have two to four times the activity, and NAA has two times the activity of IAA. Although 2,4-D, 2,4,5-T, PCPA, and Picloram are often used to induce rapid cell proliferation, exposure to high levels or prolonged exposure to these auxins, particularly 2,4-D, results in suppressed morphogeneic activity. Auxins are generally included in a culture medium to stimulate callus production and cell growth, to initiate shoots, particularly roots, and to induce somatic embryogenesis and stimulate growth from shoot apices and shoot tip cultures.

The cytokinins commonly used in the culture media include 6-benzylaminopurine or 6 benzyladenine (BAP, BA), 6-γ-γ-dimethylaminopurine (2iP), N-(2-furanylmethyl)-1H-puring-6-amine (kinetin), and 6 (4-hydroxy-3-mehty-trans-2-butenylamino) purine (zeatin). Zeatin and 2iP are considered to be naturally occurring cytokinins, while BA and kinetin are synthetically derived cytokinins. Adenine, another naturally occurring compound, has a base structure similar to that of the cytokinins and has shown cytokinin-like activity in some cases. Many plant tissues have an absolute requirement for a specific cytokinin for morphogenesis to occur, whereas some tissues are considered to be cytokinin independent, i.e., no cytokinin or a specific cytokinin may be required for organogenesis.

The cytokinins are generally added to a culture medium to stimulate cell division, to induce shoot formation and axillary shoot proliferation, and to inhibit root formation. The type of morphogenesis that occurs in a plant tissue culture largely depends upon the ratio and concentrations of auxins and cytokinins present in the medium. Root initiation of plantlets, embryogenesis, and callus initiation all generally occur when the ration of auxin to cytokinin is high, whereas adventitious and axillary shoot proliferations occur when the ration is low. The concentrations of auxins and cytokinins are equally as important as their ratio.

Gibberellins (GA_3) and abscisic acid (ABA) are two other growth regulators occasionally used in culture media. Plant tissue cultures can usually be induced to grow without either GA_3 or ABA, although, certain species may require these hormones for enhanced growth. Generally, GA_3 is added to culture media to promote the growth of low-density cell cultures, to enhance callus growth, and to elongate dwarfed or stunted plantlets. Abscisic acid is generally added to culture media to either inhibit or stimulate callus growth (depending upon the species), to enhance, inhibit, or stimulate callus growth (depending upon the species), to enhance shoot or bud proliferation, and to inhibit latter stages of embryo development.

Stock Solutions

The use of stock solutions reduces the number of repetitive operations involved in media preparation and, hence, the chance of human or experimental error. Moreover direct weighing of media components (e.g., micronutrients and hormones) that are required only in milligram or microgram quantities in the final formulation cannot be performed with sufficient accuracy for tissue culture work. For these components, preparation of concentrated stock solutions and subsequent dilution into the final media is standard procedure. In addition, concentrated solutions of some materials are more stable and can be stored for longer periods than more dilute solutions.

To prepare a stock solution, weigh out the required amount of the compound and place it in a clean flask. It is common practice to make a stock solution 10x or 100x, depending upon the solubility of the compound. Once the chemical is in the flask, dissolved it in a small amount of water, ethyl alcohol, 1 N NaOH, or 1 N HCL. Next, slowly add double-distilled water to the flask, while agitating. Continue this until the proper volume is reached. Label the flask with the name of the solution, preparation and expiration dates, and the name of the person who prepared the solution. Certain items, e.g., IAA, must be prepared and stored in amber bottles to prevent photodecomposition.

Macronutrients: Stock solutions of macronutrients can be prepared at 10 times the concentration of the final medium. A separate stock solution for calcium salts may be required to prevent precipitation. Stock solution of macronutrients can be stored safely for several weeks in a refrigerator at 2°-4 °C.

Micronutrients: Micronutrient stock solutions are generally made up at 100 times their final strength. It is recommended that micronutrient stocks be stored in either a refrigerator or freezer until needed. Micronutrient stock solutions could be stored in a refrigerator for up to 1 year without appreciable deterioration. Iron stock solutions should be prepared and stored separately from other micronutrients in an amber storage bottle. Formulations for preparing stock solutions of iron are presented later.

Vitamins: Vitamins are prepared as 100X or 1000X stock solutions and stored in a freezer (-20 °C) until used. Vitamin stock solutions should be made up each time media is prepared if a refrigerator or freezer is not available. Vitamin stock solutions can be stored safely in a refrigerator for 2-3 months but should be discarded after that time.

Growth Regulators: The auxins NAA and 2,4-D are considered to be stable and can be stored at 4°C for several months; IAA should be stored at -20 °C. Auxin stock solutions are generally prepared at 100- 1000 times the final desired concentrations. Solution of NAA and 2,4-D can be stored for several months in a refrigerator or indefinitely at -20 °C. Generally IAA and 2,4-D are dissolved in a small volume of 95% ethyl alcohol or KOH and then brought to volume with double-distilled water; NAA can be dissolved in a small amount of 1 N NaOH or KOH, which also can be used to dissolve 2, 4-D and IAA.

The cytokinins are considered to be stable and can be stored at -20 °C. Cytokinin stock solutions are generally prepared at 100X to 1000X concentrations. Many of the cytokinins are difficult to dissolve, and a few drops of either 1 N HCL, 1 N NaOH, in KOH or DMSO, are required to bring them into solution.

Storage of Stock Solutions: Storage conditions for most stock solutions have already been pointed out; however, some additional points can be made. For convenience, many labs prepare stock solutions and then divide them into aliquots sufficient to prepare from 1 to 10 liter of medium; these aliquots are stored in small vials or plastic bags in a freezer. This procedure removes the inconvenience of having to un-thaw a large volume of frozen stock each time medium is prepared. Some have found that heating in a microwave oven is a satisfactory and quick method of thawing concentrated medium.

PLANT PROPAGATION

Plant propagation is the process which grows new plants from a variety of sources: Seeds, cuttings, and other plant parts. Plant propagation can also refer to the man-made or natural dispersal of seeds.

Gentian seedlings in a plant nursery.

Sexual Propagation

One way to propagate an avocado seed.

Seeds and spores can be used for reproduction (through e.g. sowing). Seeds are typically produced from sexual reproduction within a species, because genetic recombination has occurred. A plant grown from seeds may have different characteristics from its parents. Some species produce seeds that require special conditions to germinate, such as cold treatment. The seeds of many Australian plants and plants from southern Africa and the American west require smoke or fire to germinate. Some plant species, including many trees, do not produce seeds until they reach maturity, which may take many years. Seeds can be difficult to acquire and some plants do not produce seed at all. Some plants (like certain plants modified using genetic use restriction technology) may produce seed, but not fertile seed. In certain cases, this is done to prevent the accidental spreading of these plants, for example by birds and other animal.

Asexual Propagation

Rose cuttings under plastic bottle greenhouse.

Softwood stemcuttings rooting in a controlled environment.

Plants have a number of mechanisms for asexual or vegetative reproduction. Some of these have been taken advantage of by horticulturists and gardeners to multiply or clone plants rapidly. Humans may utilize these processes as propagation methods, such as tissue culture and grafting. Plants are produced using material from a single parent and as such there is no exchange of genetic material, therefore vegetative propagation methods almost always produce plants that are identical to the parent. Vegetative reproduction uses plants parts such as roots, stems and leaves.

In some plants seeds can be produced without fertilization and the seeds contain only the genetic material of the parent plant. Therefore, propagation via asexual seeds or apomixis is asexual reproduction but not vegetative propagation.

Techniques for vegetative propagation include:

- Air or ground layering.
- Division.
- Grafting and bud grafting, widely used in fruit tree propagation.

- Micropropagation.
- Stolons or runners.
- Storage organs such as bulbs, corms, tubers and rhizomes.
- Striking or cuttings.
- Twin-scaling.
- Offsets.

Heated Propagator

A heated propagator is a horticultural device to maintain a warm and damp environment for seeds and cuttings to grow in.

This can be in the form of a clear enclosed bin sitting over a hotpad, or even a portable heater pointed at the bin. The key is to keep the moisture in the clear bin, while keeping lighting over the top of it, usually.

Seed-propagation Mat

An electric seed-propagation mat is a heated rubber mat covered by a metal cage which is used in gardening. The mats are made so that planters containing seedlings can be placed on top of the metal cage without the risk of starting a fire. In extreme cold, gardeners place a loose plastic cover over the planters/mats which creates a sort of miniature greenhouse. The constant and predictable heat allows people to garden in the winter months when the weather is generally too cold for seedlings to survive naturally. When combined with a lighting system, many plants can be grown indoors using these mats.

SOMATIC EMBRYOGENESIS

Somatic embryogenesis is an artificial process in which a plant or embryo is derived from a single somatic cell. Somatic embryos are formed from plant cells that are not normally involved in the development of embryos, i.e. ordinary plant tissue. No endosperm or seed coat is formed around a somatic embryo.

Cells derived from competent source tissue are cultured to form an undifferentiated mass of cells called a callus. Plant growth regulators in the tissue culture medium can be manipulated to induce callus formation and subsequently changed to induce embryos to form the callus. The ratio of different plant growth regulators required to induce callus or embryo formation varies with the type of plant. Somatic embryos are mainly produced in vitro and for laboratory purposes, using either solid or liquid nutrient media which contain plant growth regulators (PGR's). The main PGRs used are auxins but can contain cytokinin in a smaller amount. Shoots and roots are monopolar while somatic embryos are bipolar, allowing them to form a whole plant without culturing on multiple media types. Somatic embryogenesis has served as a model to understand the physiological and biochemical events that occur during plant developmental processes as well as a component to biotechnological advancement. The first documentation of somatic embryogenesis was by Steward et al. in 1958 and Reinert in 1959 with carrot cell suspension cultures.

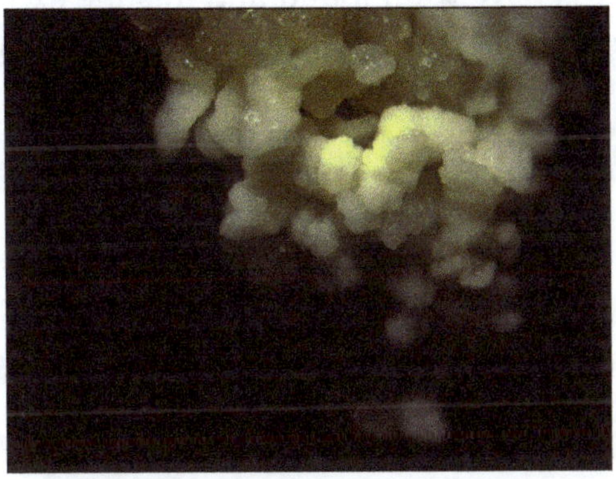

Switchgrass somatic embryos.

Direct and Indirect Embryogenesis

Somatic embryogenesis has been described to occur in two ways: Directly or indirectly:

- Direct embryogenesis: Occurs when embryos are started directly from explant tissue creating an identical clone.
- Indirect embryogenesis: Occurs when explants produced undifferentiated, or partially differentiated, cells (often referred to as callus) which then is maintained or differentiated into plant tissues such as leaf, stem, or roots. 2, 4-Dichlorophenoxyacetic acid (2, 4-D), 6-Benzylaminopurine (BAP) and Gibberellic acid (GA) has been used for development of indirect somatic embryos in strawberry (Fragaria ananassa).

Plant Regeneration by Somatic Embryogenesis

Plant regeneration via somatic embryogenesis occurs in five steps: Initiation of embryogenic cultures, proliferation of embryogenic cultures, prematuration of somatic embryos, maturation of somatic embryos and plant development on nonspecific media. Initiation and proliferation occur on a medium rich in auxin, which induces differentiation of localized meristematic cells. The auxin typically used is 2, 4-D. Once transferred to a medium with low or no auxin, these cells can then develop into mature embryos. Germination of the somatic embryo can only occur when it is mature enough to have functional root and shoot apices.

Factors Influencing

Factors and mechanisms controlling cell differentiation in somatic embryos are relatively ambiguous. Certain compounds excreted by plant tissue cultures and found in culture media have been shown necessary to coordinate cell division and morphological changes. These compounds have been identified by Chung et al. as various polysaccharides, amino acids, growth regulators, vitamins, low molecular weight compounds and polypeptides. Several signaling molecules known to influence or control the formation of somatic embryos have been found and include extracellular proteins, arabinogalactan proteins and lipochitooligosaccharides. Temperature and lighting can also affect the maturation of the somatic embryo.

Applications

Applications of this process include: Clonal propagation of genetically uniform plant material; elimination of viruses; provision of source tissue for genetic transformation; generation of whole plants from single cells called protoplasts; development of synthetic seed technology.

Uses of Somatic Embryogenesis

- Plant transformations.
- Mass propagation.

Forestry Related Example

The development of somatic embryogenesis procedures has given rise to research on seed storage proteins (SSPs) of woody plants for tree species of commercial importance, i.e., mainly gymnosperms, including white spruce. In this area of study, SSPs are used as markers to determine the embryogenic potential and competency of the embryogenic system to produce a somatic embryo biochemically similar to its zygotic counterpart.

Grossnickle et al. compared interior spruce seedlings with emblings during nursery development and through a stock quality assessment program immediately before field outplanting. Seedling shoot height, root collar diameter, and dry weight increased at a greater rate in seedlings than in emblings during the first half of the first growing season, but thereafter shoot growth was similar among all plants. By the end of the growing season, seedlings were 70% taller than emblings, had greater root collar diameter, and greater shoot dry weight. Root dry weight increased more rapidly in seedlings than in emblings during the early growing season.

During fall acclimation, the pattern of increasing dormancy release index and increasing tolerance to freezing was similar in both seedlings and emblings. Root growth capacity decreased then increased during fall acclimation, with the increase being greater in seedlings.

Assessment of stock quality just prior to planting showed that: Emblings had greater water use efficiency with decreasing predawn shoot water potential compared with seedlings; seedlings and emblings had similar water movement capability at both high and low root temperatures; net photosynthesis and needle conductance at low root temperatures were greater in seedlings than in emblings; and seedlings had greater root growth than emblings at 22 °C root, but root growth among all plants was low at 7.5 °C root temperature.

Growth and survival of interior spruce 313B Styroblock seedlings and emblings after outplanting on a reforestation site were determined by Grossnickle and Major. For both seedlings and emblings, osmotic potential at saturation (ψ_{sat}) and turgor loss point (ψ_{tip}) increased from a low of -1.82 and -2.22 MPa, respectively, just prior to planting to a seasonal high of -1.09 and -1.21 MPa, respectively, during active shoot elongation. Thereafter, seedlings and emblings (ψ_{sat}) and (ψ_{tip}) declined to -2.00 and -2.45 MPa, respectively, at the end of the growing season, which coincided with the steady decline in site temperatures and a cessation of height growth. In general, seedlings and emblings had similar ψ_{sat} and ψ_{tip} values through the growing season, and also had similar shifts in seasonal patterns of maximum modulus of elasticity, sympalstic fraction, and relative water content at turgor loss point.

Grossnickle and Major found that year-old and current-year needles of both seedlings and emblings had a similar decline in needle conductance with increasing vapour pressure deficit. Response surface models of current-year needles net photosynthesis (P_n) response to vapour pressure deficit (VPD) and photosynthetically active radiation (PAR) showed that emblings had 15% greater P_n at VPD of less than 3.0 kPa and PAR greater than 1000 µmol $m^{-2}s^{-1}$. Year-old and current-year needles of seedlings and emblings showed similar patterns of water use efficiency.

Rates of shoot growth in seedlings and emblings through the growing season were also similar to one another. Seedlings had larger shoot systems both at the time of planting and at the end of the growing season. Seedlings also had greater root development than emblings through the growing season, but root: Shoot ratios for the 2 stock types were similar at the end of the growing season, when the survival rates for seedlings and emblings were 96% and 99%, respectively.

Tracking and Fate Maps

Understanding the formation of a somatic embryo through establishment of morphological and molecular markers is important for construction of a fate map. The fate map is the foundation in which to build further research and experimentation. Two methods exist to construct a fate map: synchronous cell-division and time-lapse tracking. The latter typically works more consistently because of cell-cycle-altering chemicals and centrifuging involved in synchronous cell-division.

Angiosperms

Embryo development in angiosperms is divided into several steps. The zygote is divided asymmetrically forming a small apical cell and large basal cell. The organizational pattern is formed in the globular stage and the embryo then transitions to the cotyledonary stage. Embryo development differs in monocots and dicots. Dicots pass through the globular, heart-shaped, and torpedo stages while monocots pass through globular, scutellar, and coleoptilar stages.

Many culture systems induce and maintain somatic embryogenesis by continuous exposure to 2, 4-dichlorophenoxyacetic acid. Abscisic acid has been reported to induce somatic embryogenesis in seedlings. After callus formation, culturing on a low auxin or hormone free media will promote somatic embryo growth and root formation. In monocots, embryogenic capability is usually restricted to tissues with embryogenic or meristematic origin. Somatic cells of monocots differentiate quickly and then lose mitotic and morphogenic capability. Differences of auxin sensitivity in embryogenic callus growth between different genotypes of the same species show how variable auxin responses can be.

Carrot Daucus carota was the first and most understood species with regard to developmental pathways and molecular mechanisms. Time-lapse tracking by Toonen et al. showed that morphology of competent cells can vary based on shape and cytoplasm density. Five types of cells were identified from embryonic suspension: spherical cytoplasm-rich, spherical vacuolated, oval vacuolated, elongated vacuolated, and irregular shaped cells. Each type of cell multiplied with certain geometric symmetry. They developed into symmetrical, asymmetrical, and aberrantly-shaped cell clusters that eventually formed embryos at different frequencies. This indicates that organized growth polarity do not always exist in somatic embryogenesis.

Gymnosperms

Embryo development in gymnosperms occurs in three phases. Proembryogeny includes all stages prior to suspensor elongation. Early embryogeny includes all stages after suspensor elongation but before root meristem development. Late embryogeny includes development of root and shoot meristems. Time-lapse tracking in Norway Spruce Picea abies revealed that neither single cytoplasmic-rich cells nor vacuolated cells developed into embryos. Proembryogenic masses (PEMs), an intermediate between unorganized cells and an embryo composed of cytoplasmic-rich cells next to a vacuolated cell, are stimulated with auxin and cytokinin. Gradual removal of auxin and cytokinin and introduction of abscisic acid (ABA) will allow an embryo to form. Using somatic embryogenesis has been considered for mass production of vegetatively propagated conifer clones and cryopreservation of germplasm. However, the use of this technology for reforestation and tree breeding of conifers is in its infancy.

References

- Plant-tissue-culture-current-status-and-opportunities, recent-advances-in-plant-in-vitro-culture: intechopen.com, Retrieved 12 April, 2020
- Sahoo, Jyoti Prakash (2018-06-11). "Organogenesis and somatic embryogenesis". doi:10.13140/rg.2.2.26278.57928
- ComponentsOfTissueCultureMedia: phytotechlab.com, Retrieved 25 August, 2020
- Clapham; et al. "Gene Transfer by Particle Bombardment to Embryogenic Cultures of Picea abies and the Production of Transgenic Plantlets". Scandinavian Journal of Forest Research. 15: 151–160. doi:10.1080/028275800750014957

3

Plant Metabolic Engineering

Plant metabolic engineering redirects the enzymatic reactions for producing new compounds or enhancing the functioning of existing compounds in an organism. Primary and secondary metabolic pathways are studied under its domain. This chapter discusses the concepts associated with plant metabolic engineering in detail.

Metabolic engineering is generally defined as the redirection of one or more enzymatic reactions to produce new compounds in an organism, improve the production of existing compounds, or mediate the degradation of compounds. In highlighting progress in plant metabolic engineering over the past 25 years, it is first important to stress that it is in fact quite a young science in plants. Our knowledge of substrate-product relationships in plant pathways was reasonably well advanced by 1975 as a result of the application of radiolabel tracer studies during the previous decades. Attempts to use this knowledge to engineer metabolism in plants, however, first required the development of basic molecular biological technologies such as cloning, promoter analysis, protein targeting, plant transformation, biochemical genetics, and other areas of plant biology. Despite this delay significant progress has been made since the mid-1980s in the molecular dissection of many plant pathways and the use of cloned genes to engineer plant metabolism. Although there are numerous success stories, there has been an even greater number of studies that have yielded completely unanticipated results. Such data underscore the fragmented state of our understanding of plant metabolism and highlight the growing gap between our ability to clone, study and manipulate individual genes and proteins and our understanding of how they are integrated into and impact the complex metabolic networks in plants. With an estimated 100,000 unique compounds produced in the plant kingdom, elucidating these metabolic networks is likely to be an exciting endeavor.

Technological Advancements in Gene Discovery

The dependence of progress in plant metabolic engineering upon technological advancements is beautifully exemplified by research in lipid metabolism. Though much was learned at the biochemical level about individual steps of plant lipid synthesis during the 1970s and early 1980s, progress in purifying and cloning many pathway enzymes, especially those that are membrane associated, was hindered due to biochemical difficulties inherent in the target enzymes. A major breakthrough in the field came from genetic dissection of the pathway in Arabidopsis. This pioneering work in plant biochemical genetics was modeled after mutation-based approaches to study metabolism in bacterial systems. Earlier work by Somerville and coworkers studying photorespiration in Arabidopsis proved the feasibility of using biochemical genetics to dissect plant pathways. For lipid biosynthesis, over 10,000 mutated Arabidopsis plants were screened by gas chromatography for altered fatty acid profiles. This resulted in identification of a suite of novel and informative mutations defining steps of the lipid biosynthetic pathway, which allowed genetic models of the plastidic and extraplastidic pathways to be developed and tested. The same mutants provided genetic targets for subsequent cloning of several pathway genes by chromosome walking, T-DNA tagging, and various homology-based screening approaches. Our understanding of the biosynthesis of other classes of plant compounds such as amino acids, waxes, anthocyanins, and ascorbic acid has been similarly advanced by analogous molecular genetic approaches dissecting their respective pathways.

A second example of technology's impact on gene discovery for plant metabolic engineering comes from work on the carotenoid biosynthetic pathway. Though the pathway in plants had been known since the mid 1960s, the labile, membrane-associated enzymes remained recalcitrant to isolation and study. However, because carotenoids are also synthesized by many photosynthetic and non-photosynthetic bacteria, the development of molecular genetic tools in prokaryotes during the 1980s allowed plant researchers to access carotenoid biosynthetic genes from prokaryotes. Integrating prokaryotic systems into their work enabled researchers to finally clone the majority of carotenoid biosynthetic enzymes from plant during the 1990s.

One general approach used mutant complementation to identify and isolate carotenoid biosynthetic genes based on their resistance to specific herbicides. An early success was the cloning of phytoene desaturase (PDS) from the cyanobacterium SynechococcusPCC7942. Mutants resistant to an herbicide that inhibits PDS activity were first selected, and a library of the mutant DNA were transformed into wild-type Synechococcus.The mutant PDS gene was identified by its ability to confer herbicide resistance in the wild-type background. The evolutionary relationship of cyanobacteria and plants quickly allowed isolation of PDS orthologs from a number of plant species. A second approach, termed color complementation, engineered carotenoid biosynthetic genes from bacterial, fungal, and plant sources onto a single plasmid for expression in Escherichia coli, which normally lacks endogenous carotenoids and the associated enzymes. Depending on which genes and portion of the pathway were engineered, accumulation of variously colored pathway intermediates resulted. Transformation of plant cDNA expression libraries into such E. colibackgrounds allowed functional identification of the rare cDNAs (often one in several hundred thousand) encoding the next enzyme of the pathway based on the associated change in color of the carotenoid product. Similar strategies utilizing heterologous systems have also allowed function-based cloning of enzymes for the synthesis of plant sterols, amino acids, and vitamins.

Predicting the Outcome of Metabolic Engineering is a Challenging Job

Although progress in pathway gene discovery and our ability to manipulate gene expression in transgenic plants has been most impressive during the past two decades, attempts to use these tools to engineer plant metabolism has met with more limited success. Though there are notable exceptions, most attempts at metabolic engineering have focused on modifying (positively or negatively) the expression of single genes affecting pathways. In general, the ability to predict experimental outcomes has been much better when one is targeting conversion or modification of an existing compound to another rather than attempting to increase flux through a pathway. Modifications to metabolic storage products or secondary metabolic pathways, which often have relatively flexible roles in plant biology, have also been generally more successful than manipulations of primary and intermediary metabolism.

As was the case for gene discovery, the lipid biosynthetic pathway was one of the earlier pathways to be targeted for manipulation and represents one of the better examples of metabolic engineering in plants to date. Most enzymes for fatty acid synthesis in plants have been cloned and various academic and industrial groups have modified their expression to manipulate oilseed fatty acid composition. The engineering of soybean and canola to produce higher levels of mono-unsaturated fatty acids was undertaken because their oils contain high levels of linolenic acid, which is susceptible to oxidation and limits the shelf life and utility of these oils. Antisense inhibition of oleate desaturase expression resulted in oil that contained >80% oleic acid (a mono-unsaturated fatty acid) and had a significant decrease in polyunsaturated fatty acids. The resulting mono-unsaturate-rich oils are more stable to oxidation, healthier in the human diet than the corresponding poly-unsaturate containing oils, and represent an excellent early example of the practical application of metabolic engineering in plants.

Other areas of plant metabolism with high potential to benefit human health have also been successfully engineered in recent years. In one example, the last enzyme in the synthesis of α-tocopherol, γ-tocopherol methyltransferase (γ-TMT) was used to increase the vitamin E activity of Arabidopsis seed oil. Arabidopsis seed, like most oilseed crops, contains a high proportion of γ-tocopherol, which has 10% of the vitamin E activity of α-tocopherol. Expression of γ-TMT in Arabidopsis seed resulted in the conversion of the large pool of γ-tocopherol to α-tocopherol with a corresponding 10-fold increase in vitamin E activity. Engineering similar conversions in soybean, canola and maize would elevate the levels of this important antioxidant/vitamin in the diet and potentially have significant health consequences for the general population. In an example of metabolic engineering of plant vitamin content targeted at the developing world, β-carotene (provitamin A) was recently engineered into rice endosperm. Vitamin A deficiency is a serious health issue in many parts of the developing world. Rice is a major staple in developing countries, but is a poor source of many essential vitamins and minerals, including β-carotene (provitamin A). Efforts to engineer β-carotene production in rice benefited directly from the identification of carotenoid biosynthetic genes in model systems described earlier. Three carotenoid biosynthetic enzymes (two from plants and one from bacteria) were engineered for simultaneous expression in rice endosperm. The resulting first generation transgenic rice produced yellow endosperm (so-called "golden rice") containing β-carotene at levels that would provide 10% of the recommended daily allowance with an average daily rice intake. Subsequent manipulation may allow the vitamin A recommended daily allowance to be approached with a daily rice intake and potentially provide relief from vitamin A deficiency for millions worldwide.

When faced with the novel experimental possibilities that molecular, genomic, and transgenic approaches have presented over the past two decades, researchers can be tempted to become fixated on producing transgenic plants and lose appreciation for the important roles enzyme kinetics play at individual reaction steps and within entire pathways. The results of careful consideration of enzyme kinetics in metabolic engineering were elegantly demonstrated in research directed at modifying starch synthesis by manipulating ADP-Glc pyrophosphorylase (ADPGPP). Plant ADPGPP is sensitive to allosteric effectors and has been proposed to be a key regulator limiting starch synthesis. Escherichia coli ADPGPP is involved in glycogen synthesis and is also sensitive to allosteric effectors. Mutations affecting allosteric regulation cause an increase in glycogen levels in E. coli. Stark et al. engineered wild-type and mutant E. coli ADPGPP for expression in plants and assayed the effect on starch accumulation. Tubers from potato plants transformed with the wild-type E. coli enzyme had starch levels similar to wild-type plants, whereas those transformed with the allosterically insensitive E. coli ADPGPP enzyme had starch levels up to 60% higher than wild type. The effect was only observed when the mutant protein was targeted to the chloroplast and driven by a tuber specific promoter; constitutive expression was lethal. Such results demonstrate the importance of considering the target tissues, subcellular localization, and kinetics of enzymes when engineering plant metabolism.

Attempts to manipulate the Lys content of seeds (Lys is a limiting amino acid in most seeds used for food or feed) illustrate that one needs to consider catabolic, as well as anabolic, variables when trying to engineer a particular metabolic phenotype in plants. A key step in Lys synthesis is carried out by dihydrodipicolinate synthase, which is feedback inhibited by the pathway end product, Lys, and thus plays a key role in regulating flux through the pathway. Engineering plants to overexpress a feedback insensitive bacterial dihydrodipicolinate synthase, similar to the approach with ADPGPP described earlier, greatly increased flux through the Lys biosynthetic pathway. However, in most cases this did not result in increased steady-state Lys levels as the plants also responded by increasing flux through the Lys catabolic pathway. Substantial increases in Lys only occurred in plants where flux increased to such a level that the first enzyme of the catabolic pathway became saturated.

The manipulation of well-characterized "rate-limiting" enzymes of primary carbon metabolism to study their role in regulating pathway flux has provided some of the more surprising results from metabolic engineering in plants. These experiments drive home the point that a thorough understanding of the individual kinetic properties of enzymes may not be informative as to their role in complex metabolic pathways. Potential regulatory enzymes are generally identified based on their catalyzing irreversible reactions and being regulated by appropriate effector molecules for a pathway; traditional biochemical hallmarks of rate controlling enzymes. When the highly regulated Calvin cycle enzymes Fru-1, 6-bisphosphatase and phoshphoribulokinase were reduced 3- and 10-fold in activity, respectively, surprisingly minor effects were observed on the photosynthetic rate. In contrast, a minor degree of inhibition of plastid aldolase, which catalyzes a reversible reaction and is not subject to allosteric regulation, led to significant decreases in photosynthetic rate and carbon partitioning. Thus aldolase, an enzyme seemingly irrelevant in regulating pathway flux, was shown to have a major control over the pathway. Analogous surprises were also found when manipulating presumed "rate limiting" enzymes of glycolysis. Such data has called into question many of the longstanding ideas about flux regulation in plants and is forcing a reassessment of the role of individual enzymes in the process. These studies also make clear the caution that must be exercised when extrapolating individual enzyme kinetics to the control of pathway flux.

Transcriptional Regulators may allow Manipulation of Entire Pathways

An intriguing approach for metabolic engineering and increasing our understanding of the coordinate changes in gene expression needed to regulate entire pathways is to identify and study transcriptional factors controlling pathways or branches of metabolism. Many of the transcriptional regulators affecting plant biochemistry and development were originally identified by chemical- or transposon-based mutant screens in maize, snapdragon or Arabidopsis. The cloning of such loci has provided the opportunity to use these genes to manipulate plant biochemistry in the host organism or in other plants. One of the early instances of using this approach to manipulate plant biochemistry was the engineering of Arabidopsis to express the maize transcription factors C1 and R, which regulate production of anthocyanins in maize aleurone layers. Expression of C1 and R together from a strong promoter caused massive accumulation of anthocyanins in Arabidopsis, presumably by activating the entire pathway. More recently, the maize transcriptional regulators C1, R, and P were expressed in cell cultures and the effect on anthocyanin biochemistry and global gene expression analyzed. Novel insights into the anthocyanin pathway, its regulation, and additional differentially expressed targets of these regulatory genes were obtained. Such expression experiments hold great promise and may eventually allow the determination of transcriptional regulatory networks for biochemical pathways.

KEY APPLICATIONS OF PLANT METABOLIC ENGINEERING

Plants constitute a remarkable feat of metabolic engineering. Not only does their energy derive entirely from the sun and their carbon from CO_2, but they can defend themselves from pests and predators without the benefit of mobility; they participate in complex symbioses, in part by tailoring the composition of their epi- and endophytic microbial communities, and they can survive extremes of temperature and nutrient and water availability. What more could we ask of plants?

A great deal, it turns out. Both conventional breeding and modern metabolic engineering have been used to boost productivity and to enhance fitness (for example, by increased resistance to pests, herbicides, and climatic extremes). In addition, new areas of application have been introduced that would have seemed like science fiction only a few decades ago, including the use of plants to produce vaccines, bioplastics, and derivatives of complex natural product drugs. Many of these more recent engineering goals could not have been accomplished by fine-tuning endogenous host metabolism; instead, they required the installation of new metabolic pathways from other plants or bacteria. Adding new nodes to a plant metabolic network is a difficult task that will benefit from advances in targeted genome modification, tissue, cell, and organelle-specific gene expression, and the controlled expression of multi-gene pathways.

Plants that can Fix their own Nitrogen

A remarkable 180 million tons of nitrogen fertilizer is used every year in industrial farming, and it has played such an enabling role that fertilizer use is estimated to meet the nutritional needs of one-third of the human population. However, the use of nitrogen fertilizers has serious disadvantages, including substantial cost and deleterious effects on the soil and surrounding environment, and neighboring waterways. Plants that could satisfy some or all of their own nitrogen requirement would transform agriculture by reducing or eliminating this enormous dependence on fertilizer.

There are two conceivable ways in which a plant could be engineered to satisfy its own nitrogen requirement. The first takes advantage of the fact that some bacteria carry out their own version of the Haber-Bosch process—reducing atmospheric N_2 into a more bioavailable form, NH_3—using the enzyme nitrogenase. This complex enzyme contains multiple metalloclusters and requires a large quantity of biochemical energy to transfer the electrons needed to activate the exceptionally stable N_2 triple bond. By expressing nitrogenase, plants would be able to fix their own nitrogen. The primary advantage of this approach is that it is direct: nitrogen fixed by a plant could be used immediately to generate amino acid and nucleic acid monomers and transport them to neighboring cells. Although the process would incur a metabolic cost, it could be regulated by the endogenous level of nitrogen to maximize its efficiency.

Immense technological challenges stand in the way of accomplishing this goal. Eighteen gene products (in Klebsiella oxytoca) are necessary and sufficient for the production of nitrogenase and its complex iron-molybdenum cofactor. By an impressive feat of microbial engineering, the biosynthetic gene cluster for nitrogenase has been refactored—taken apart, recoded, and put back together using known components—and shown to be active in its new host. The successful transfer of other large gene clusters from one microbe to another, such as the one encoding the magnetosome, suggests that the process of functionalizing microbes is undergoing a dramatic improvement.

But new challenges must be overcome for the expression of similar elements in plants. First, all 18 components of the nitrogenase biosynthetic apparatus would need to be expressed simultaneously in plants and function in concert, a considerable barrier given that the largest number of genes expressed in an engineered plant to date is eight in the establishment of a photorespiratory bypass in Arabidopsis. Second, since plants are eukaryotic and multicellular, where in the plant cell should the genes be expressed and in which cell types of the plant? This question is especially relevant for nitrogenase, which is poisoned by oxygen and must therefore be expressed under anaerobic conditions. Tools that enable organelle- and cell-type specific expression will be of great utility here and in other plant engineering efforts.

The second way to reduce or eliminate the need for nitrogen fertilizer would be to engineer a rhizosphere symbiosis between a nitrogen-fixing microorganism and a plant host. Although this approach is less direct than expressing nitrogenase in plants, it has two primary advantages: (1) It uncouples the difficulties of utilizing nitrogenase (e.g., sequestering the enzyme in an anaerobic compartment) from the biology of the plant host, and outsources the demanding chemistry involved to a bacterial strain better suited to the task. The well-studied symbioses between legumes and their nitrogen-fixing, root nodulating bacterial symbionts prove that a bacterial mutualist can satisfy the nitrogen needs of a plant host. Even though root-nodulating bacteria appear to be specific to legumes, the presence of nitrogen-fixing bacteria in the rhizosphere opens the possibility that symbioses of this sort are a much more widely distributed phenomenon. If so, then the feasibility of making this mode of nitrogen exchange more efficient—rather than engineering it from scratch—would appear high. However, there remain two primary challenges in engineering a stable and practical rhizosphere symbiosis: Enabling efficient nutrient exchange and maintaining specificity of the host-microbe pair. Both could take years to develop and are likely to require not just plant but also microbial metabolic engineering. In the meantime, advanced molecular breeding tools that enable access to natural variation in a plant's wild ancestors are a promising alternative approach to increasing crop plant yields.

Crop Plants with Altered Nutrient Content

What if maize could be as nutritious as broccoli, and broccoli as palatable as maize? Golden rice proves the concept that the nutrient content of a crop plant can be improved by metabolic engineering. Centuries of breeding not just rice but maize, wheat, tomatoes, and cruciferous vegetables have undoubtedly enhanced the levels of certain nutrients while suppressing others. In the long run, adding the beta-carotene pathway to rice, to produce rice with higher levels of vitamin A (yielding the strain known as Golden rice), may prove to be among the more modest changes to nutrient content that are achievable by metabolic engineering.

One class of targets for nutrient engineering are the metabolic pathways that produce phytoalexins, flavonoids, and other molecules that are thought to play a role in the chemopreventive properties of vegetables and fruits. One approach is to alter the level of a nutrient in its native host, a strategy that does not require knowing the genes in its biosynthetic pathway. Two examples illustrate the feasibility of this approach. In the first, a tour de force in conventional plant breeding led to the Beneforte strain of broccoli, which has higher levels of the glucosinolate glucoraphanin than traditional cultivars of broccoli (and is already being sold at some US grocery stores). Early studies have demonstrated that human consumption of high-glucoraphanin broccoli results in improved metabolism and reduced levels of fatty acids and in the lipid compounds associated with inflammation. In the second example, by expressing two transcription factors from snapdragon in tomato, the levels of the flavonoid anthocyanin have been increased 3-fold, a level that was sufficient for this flavonoid to confer improved chemo preventive properties in cancer susceptible mice.

A different approach would be to express the pathway for a health-promoting molecule in a new host. Golden rice provides one example of this strategy, albeit a simple one from the viewpoint of genetic engineering since only two genes were required for the production of beta-carotene (although betacarotene is an endogenous metabolite, it is not produced in the endosperm, the edible portion of the plant). A more ambitious prospect would be to transfer the 13-gene glucoraphanin pathway to a widely consumed crop such as rice, wheat, or maize. Such an effort would require new approaches for discovering the genes for plant nutrient biosynthetic pathways. In addition, new tools for site-selective genome editing would be needed to express multiple enzymes in a pathway or multiple pathways in a plant. By analogy to previous efforts in bacterial metabolic engineering, additional genetic changes might be required to boost the levels of key precursors to the molecule of interest.

Introducing or increasing the level of a nutrient compound could also alter the taste of a plant, potentially impacting its palatability. Likewise, the challenge of improving the taste profile of a plant could be addressed by a similar strategy, since small molecule metabolites make an important contribution to flavor. Notable targets in this area include the steviol glycosides and the mixed esters that give strawberry plants their distinctive flavor.

Engineering Crops for Biofuel Production

Plants would seem to be the ideal "invention" to combat the dual challenges of rising greenhouse gases and the need for green energy: they capture CO_2 from the air and turn it into sugar, the ideal substrate for biofuel production. Not surprisingly, though, plants protect this energy rich polymer—most of the carbon is stored as dehydrated crystalline cellulose, wrapped in a meshwork of crosslinked phenylpropanoids, lignin. In solving the problem of how to create a rigid skeleton that enables a sessile organism to resist pests, pathogens, wind, precipitation, and temperature extremes, plants have managed to sequester their carbon in a form that is complicated to unwrap.

Cellulose presents a challenge in itself. Other polymers of glucose, including starch and glycogen, are rich energy sources that are easily degraded into sugar monomers. The beta-1,4-linked chains of cellulose, on the other hand, can pack tightly together, excluding water in a way that prevents glycosidic enzymes from releasing its constituent sugar monomers. On top of the challenge of hydrolyzing cellulose, the lignin with which it copurifies can inhibit the action of cellulases, creating a need for costly and energyintensive pretreatment to separate the cellulose from lignin.

Strategies have been proposed in which the lignin could be degraded into valuable aromatic monomers either chemically or by enzymes found in white rot fungi, microorganisms that naturally degrade lignin. Both remain exciting prospects, although neither has been shown to work in a real-world setting.

Alternatively, lignin biosynthesis can be genetically modified to change its chemical composition or to reduce its content in plant tissues to improve the processing of polysaccharides. For example, an Arabidopsis thaliana knockout mutant of caffeoyl shikimate esterase, a newly discovered lignin biosynthetic gene, gave a 4-fold increase in the efficiency of saccharification—the breakdown of oligosaccharides to monosaccharides—relative to wild-type. However, typical of many ligninmodified plants, there were discernible effects on plant growth and development; the transgenics contained 25% less cellulose content and were 40% lighter and smaller than wild-type. In another recent attempt, protein engineering and expression of a 4-O-methyltransferase in A. thaliana substantially reduced lignin content by blocking access to phydroxyls of lignin precursors needed for polymerization. Interestingly, no significant changes in growth phenotype were observed and saccharification yields improved by 25%. Further success with lignin modifications to plants will depend heavily on an improved understanding of lignin biosynthesis and of the physiological consequences of altering its structure.

But what if plants did the difficult work on their own, without the need for largescale reactors or any human intervention? Following the harvest of a biofuel feedstock, a crop grown for biofuel production, the ideal scenario would be for plants to degrade their own lignin, releasing pure cellulose that could be more easily degraded into glucose. For that matter, the plant could be engineered to break down its own cellulose on demand, releasing fermentation-ready sugars for biofuel production.

Two formidable challenges stand in the way of realizing this goal. The first is the feasibility of enzymatically degrading lignin to liberate cellulose. Although this would undoubtedly be a difficult task, the ability of white rot fungi to degrade lignin proves the concept that there exist enzymes (e.g., lignin peroxidases) that can cleave the lignin meshwork into monomers and smaller polymers. Nor is it necessary for the lignin to be degraded; an alternative possibility would be to more heavily crosslink (and consolidate) the lignin, causing it to precipitate and making it easier to separate from cellulose. Since this is a process that will likely be carried out by suites of degrading enzymes in rot fungi, a critical step would be to first identify sets of enzymes that could be co-expressed to make the necessary modifications to lignin.

An additional advance would be to regulate the production of such enzymes with exquisite specificity. Since transcription and translation are unlikely to persist long after harvest, the enzymes could instead be expressed in an inactive, caged form while the plant is still alive. Any small amount of leaky activity prior to harvest could compromise the plant structurally, or make it susceptible to pathogens and pests. Too little activity, on the other hand, would fail to release the cellulose from lignin. One informative model is that of myrosinases, enzymes in cruciferous plants that activate glucosinolates, a family of defense compounds, by glycoside hydrolysis. These enzymes are translated but physically sequestered from their glucosinolate substrates in different cells, activated only upon injury of the plant tissue. Physical sequestration of the constituent enzymes in a "degradome" could likewise serve to cage their activity until after harvest. A recent paper reports a promising alternative strategy based on incorporating chemically labile bonds into the lignin backbone.

Enhancing Photosynthetic Efficiency

As the world population and urbanization rapidly increase, increases in crop yield will be required to meet growing food demands. Cereals will continue to be an especially important part of the future food supply both for direct consumption and as livestock feed. Crop yields for biofuel feedstocks will also need to be considered for green energy production.

Rubisco is the enzyme that catalyzes the first key step in CO_2 fixation as part of the Calvin Cycle. Its low turnover rate and ability to also use oxygen as a substrate in photorespiration make it notoriously inefficient. As a result, plants make more Rubisco than any other protein, making it the most abundant protein in the world.

Alternative carbon fixation systems have evolved to improve the efficiency of photosynthesis by actively concentrating CO_2 and reducing the oxygenase activity of Rubisco. In plants, the two systems that have evolved to improve photosynthesis efficiency are C4 and Crassulacean acid metabolism (CAM) photosynthesis. Unlike in C3 plants (e.g., rice and other cool-season cereals), in which the process of fixing carbon into C3 acids occurs in the same cell type, C4 plants have evolved to separate the Calvin cycle and carbon capture into different cell types. CO_2 is first captured within mesophyll cells to produce C4 acids, which diffuse to bundle sheath cells where they are decarboxylated and concentrated to maximize Rubisco's carboxylating efficiency. In CAM plants, photosynthesis and carbon capture are separated temporally instead of spatially. Living in arid conditions, CAM plants capture CO_2 at night into C4 acids, enabling them to close their stomata during the day to prevent water loss. As with C4 photosynthesis, the C4 acids generated by CAM photosynthesis are decarboxylated and concentrated to enhance Rubisco's efficiency.

C4 plants generally have better nitrogen and water-use efficiencies than C3, particularly in hot and dry climates. C4 plants also have radiation use efficiencies—the ability to convert photons into biomass— that are 50% greater than those of C3 plants, which has led many to suggest that engineering C3 plants into C4 could potentially lead to a 50% increase in crop yield. The International Rice Research Institute is currently working toward incorporating C4 photosynthesis into rice.

The evolution of C3 to C4 plants has occurred independently in more than 60 different taxa, indicating that this engineering effort may be more feasible than it might otherwise appear, but many new plant metabolic engineering technologies will be required. All the enzymes of the C4 cycle are known and already exist in C3 plants. However, expressing the enzymes of the C4 cycle alone will not be enough, as the plant's anatomy is crucial for the success of the pathway. The genes that are responsible for controlling C4 leaf anatomy remain largely unknown and are being identified by mutant populations of model C4 plants like Sorghum. Additionally, cell-specific promoters will need to be identified to enable cell-type-specific expression in bundle-sheath or mesophyll cells. Finally, with >20 genes needed for the installation of C4 photosynthesis in C3 plants, the development of more sophisticated transformation and genome editing technologies will be required.

METABOLIC PATHWAYS

Plant metabolism is highly compartmented. The presence of several organelles, each performing specific physiological and metabolic roles, shows developmental plasticity, which means that their in situ metabolic pool sizes will vary within a cell and be dependent on the developmental stage. Knowing the amount of metabolites in the compartment helps to determine if its content is sufficient to participate in the metabolic process of that organelle or cell. This information is critical for accurate pathway engineering. It is not trivial to obtain quantities of relatively pure fractions of subcellular organelles adequate for analysis. Non-aqueous subcellular fractionation can be used to separate organelles and is technically easier with non-bulky tissues such as leaves, but great progress has also been made with hitherto difficult organs such as tubers. Use of this technique in combination with GC-MS revealed that most organic and amino acids and sugars are sequestered in the tuber vacuoles, but that their concentrations are highest in the cytosol. This information can be used to determine the potential activation state of enzymes in each compartment and help to develop transgenic approaches for modifying enzyme activity.

PATHWAYS AND CYCLES

Chemical reactions in the cell occur in a sequence of stages called a metabolic pathway. Each stage is catalyzed by an enzyme, a protein that changes (usually increases) the rate at which the reaction proceeds but does not alter the reactants or end products. Certain thermodynamic conditions must be met for a reaction to proceed, even in the presence of enzymes. If the end product of the reaction is also the reactant (or substrate) that starts the pathway, then the sequence of reactions is called a metabolic cycle. The intermediate chemicals that are formed and used in the various stages of the sequence are called intermediary metabolites.

Metabolic pathways and cycles are either catabolic (energy-releasing) or anabolic (energy-consuming). Catabolic reactions break down complex metabolites into simpler ones, whereas anabolic reactions build up (biosynthesize) new molecules. When chemical bonds are broken, energy is released, which drives anabolic reactions to form new bonds. The energy released generally has been stored in high-energy bonds of an intermediate energy carrier molecule, such as the terminal phosphate bond of adenosine triphosphate (ATP). (When the terminal phosphate is split from the ATP molecule, adenosine diphosphate, or ADP, is formed and inorganic phosphate is released, along with energy.) The simpler metabolites formed via catabolic reactions are often the building-block metabolites used in anabolic reactions to synthesize more complex molecules (e.g., starch, proteins, or lipids).

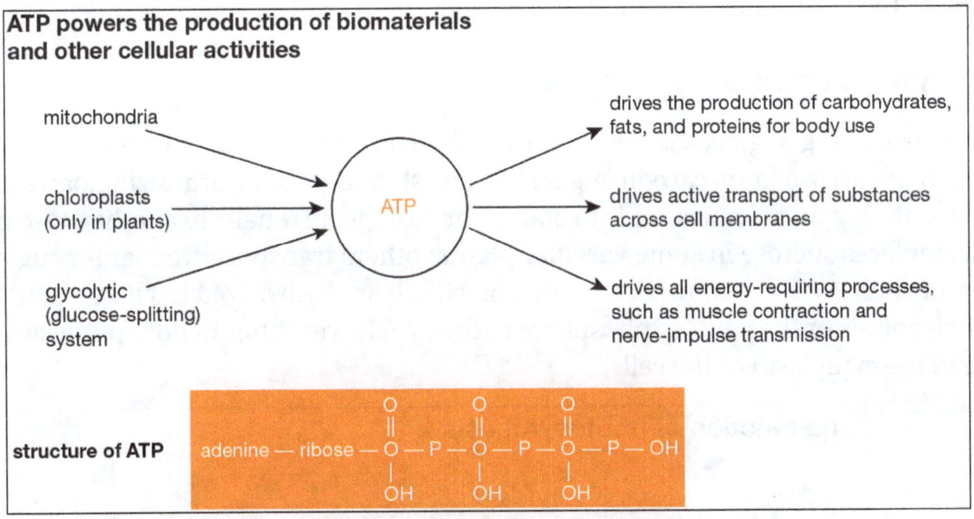

Adenosine triphosphate; physiology.

Adenosine triphosphate (ATP) is the power source of many biochemical reactions. It is produced in the cell structures and system listed at the left to energize the important life processes listed on the right. An abbreviated chemical formula of the structure of ATP is also shown. The two high-energy P–O–P bonds are responsible for its power.

Control Mechanisms

The cells of all plants are eukaryotic, because they possess a nucleus and membrane-bound organelles, such as chloroplasts, mitochondria, glyoxysomes, peroxisomes, and vacuoles. The thousands of metabolic reactions that take place in the cell are regulated within these organelles and their subcompartments. When compared with cells of other eukaryotic organisms, plant cells have a high degree of metabolic compartmentalization.

The primary mechanism of metabolic control, however, remains the enzymes themselves. Although all enzymes of the pathway help determine the net and directional flow of carbon, certain key stages are controlled by regulatory enzymes. Regulatory enzymes may either catalyze the first stage in the metabolic pathway or catalyze reactions in which key branch points occur. The activity of such enzymes, in turn, may be controlled by the amount synthesized (coarse control by gene expression)—in that further action by the enzyme is inhibited when some critical concentration of the reaction product is reached—or by special metabolites, called effectors, that interact directly with the enzyme (fine control). The latter metabolites may be either part of or totally unrelated to the metabolites of the pathway.

Another mechanism by which metabolic reactions are regulated is through transport systems in the membranes of organelles. These systems control the nature, direction, and amount of metabolites entering the metabolic pathways and are often uniquely related to the autotrophic nutrition of plants.

Principal Pathways and Cycles

The 6-carbon sugar glucose, a product of photosynthesis, is mostly translocated in the form of sucrose (a 12-carbon sugar) to nourish non-photosynthesizing parts of the plant, or it may be polymerized into starch for storage. (Trehalose, another 12-carbon sugar, replaces sucrose in some vascular plants; others transport even larger sugars or sugar alcohols). When required, sucrose and starch are hydrolyzed to glucose and then enter glycolysis or the pentose phosphate pathway. The reactions of both pathways take place in the cytoplasm of the cell.

The reaction of photosynthesis

carbon dioxide + water —(energy from light; chlorophyll in leaves)→ glucose + oxygen

$$6CO_2 + 6H_2O \longrightarrow C_6H_{12}O_6 + 6O_2$$

Photosynthesis: In photosynthesis, plants consume carbon dioxide and water and produce glucose and oxygen. Energy for this process is provided by light, which is absorbed by pigments, primarily chlorophyll. Chlorophyll is the pigment that gives plants their green colour.

The net result of glycolysis is the metabolism of glucose into two molecules of the four-carbon organic acid malate. This metabolic pathway involves phosphate-containing intermediates and is regulated by two enzymes, which catalyze those reactions that contain the substrates fructose phosphate and phosphoenolpyruvate (PEP). Glycolysis yields ATP molecules and hydrogen; the latter is accepted by the coenzyme (coenzymes are smaller, nonprotein participants associated with certain enzymes) nicotinamide adenine dinucleotide (NAD) to form NADH. The hydrogen on NADH then reacts either with molecular oxygen (O_2) to capture the energy (and transfer it to the high-energy bonds of ATP) or with another metabolite to reduce the molecule by the addition of hydrogen. Some intermediates are used in the biosyntheses of fat or certain amino acids.

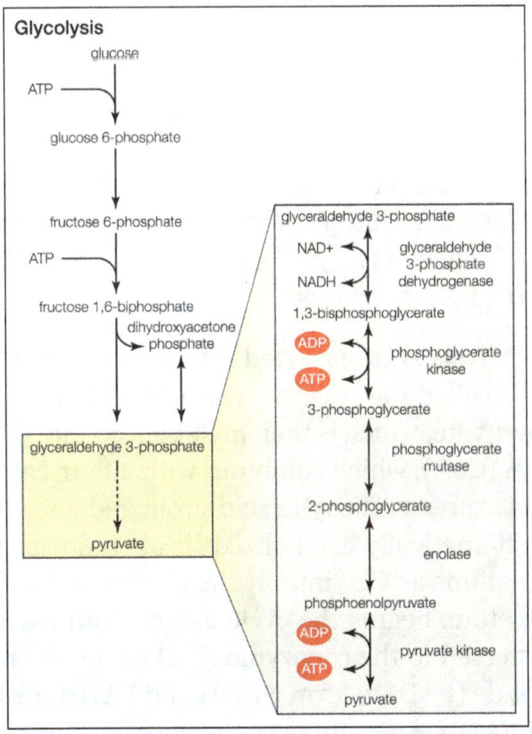

The generation of pyruvate through the process of glycolysis is the first step in fermentation.

The pentose phosphate pathway is an alternative pathway for the catabolism of glucose, producing end products that are used in the biosynthesis of nucleic acids, some vitamins, and key metabolites. It also furnishes reducing power (i.e., it accepts hydrogen atoms and carries them on the coenzyme nicotinamide adenine dinucleotide phosphate [NADP]) for use in the synthesis of substances such as fat. It is regulated by the rate at which the product of the pentose pathway, NADPH, is oxidized.

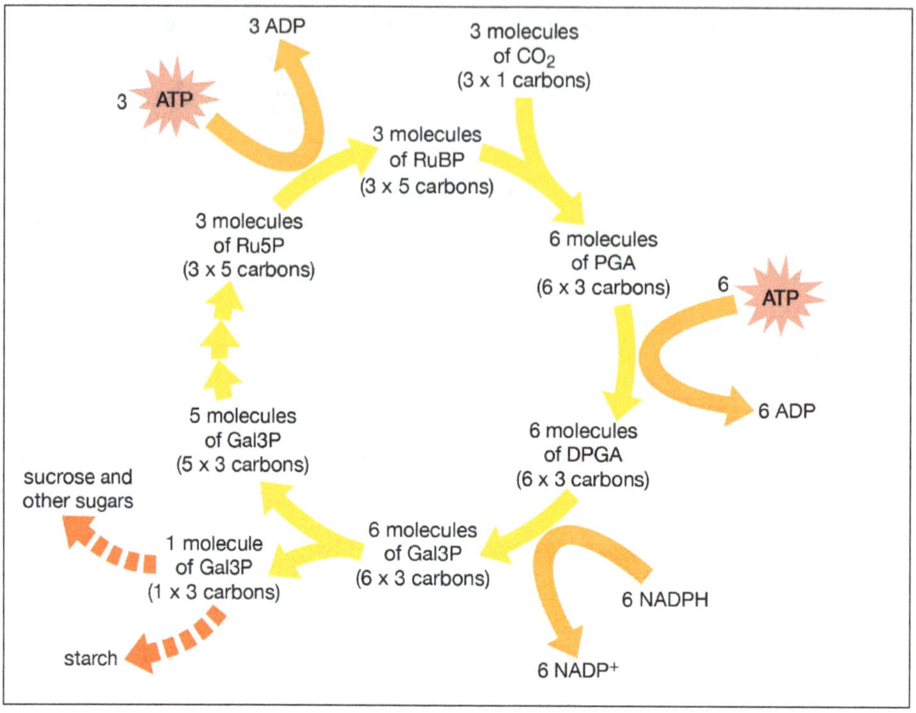

C3 carbon fixation pathway.

Pathway of carbon dioxide fixation and reduction in photosynthesis, the reductive pentose phosphate cycle. The diagram represents one complete turn of the cycle, with the net production of one molecule of Gal3P. The nine molecules of ATP and six molecules of NADPH come from the light reactions.

Malate produced in glycolysis is transported into the mitochondria, where it enters a sequence of 10 reactions called the tricarboxylic acid (TCA) cycle, or Krebs cycle. Malate is converted into pyruvate, which is then metabolized into the two-carbon intermediate, acetyl coenzyme A (CoA), which combines with a four-carbon acid, oxaloacetate. The product, citrate, has three carboxylic acid groups—hence the name tricarboxylic acid cycle. Citrate is systematically catabolized (broken down) with progressive losses of successive carbon atoms as CO_2 into five-carbon and, finally, four-carbon, acids. The latter acid, oxaloacetate, begins the cycle again. With each oxidation reaction, a hydrogen atom is transferred to the coenzyme NAD or, in one reaction, the coenzyme flavin adenine dinucleotide (FAD) to form NADH and FADH, respectively. The reduced coenzymes NADH and FADH enter into a sequence of reactions called the respiratory chain on the inner membrane of the mitochondrion. This chain is a series of carriers (ubiquinone and several iron-containing chemicals called cytochromes) that ultimately transfer the hydrogen and electrons of these coenzymes to molecular oxygen, forming water. The energy generated from the oxidation by the respiratory chain is trapped in three ATP molecules formed per NADH molecule oxidized. The mechanism is chemiosmotic in that it involves building a hydrogen ion (proton) gradient on one side of the mitochondrial membrane.

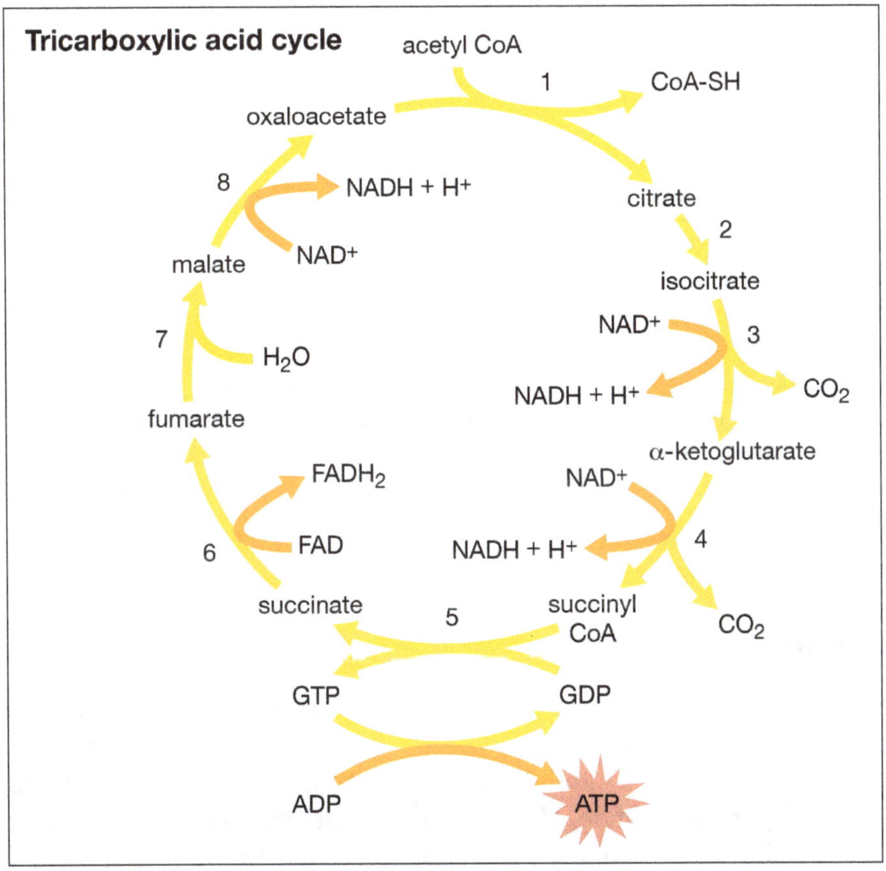

The eight-step tricarboxylic acid cycle.

A net of 36 ATP molecules are gained from all hydrogen-carrying coenzymes formed in glycolysis and the TCA cycle, and they represent the principal energy source for most anabolic (biosynthetic) reactions in plants. In addition, the TCA cycle furnishes metabolites for the biosynthesis of important organic molecules of the cell.

Another metabolic cycle, the isoprenoid pathway, produces essential oils, carotenoid pigments, certain plant hormones, and rubber. These metabolites are unique to plants and serve such functions as attracting pollinating insects, providing defense against herbivores, and producing photosynthetic pigments and phytohormones. Plant seedlings use the glyoxylic acid cycle to convert fats (principally from seeds) into glucose. This occurs initially in the glyoxysome and subsequently in the mitochondria and cytosol (the fluid mass that surrounds the various organelles).

Unique Features of Plant Metabolism

The pathways outlined above exist in essentially the same form in all organisms, but metabolism in plants does have certain unique features. Plant mitochondria, for example, have specific transport systems for the NADH produced in glycolysis and for the oxaloacetate produced from a direct fixation of CO_2 into PEP. Unlike animal mitochondria, plant mitochondria metabolize malate and the amino acid glycine. A special enzyme converts malate to pyruvate, thereby allowing an alternative to the glycolytic pathway that is common in other organisms. Glycine is a product of the unique plant pathway of photorespiration.

Plant mitochondria possess a cyanide-resistant alternative respiratory chain in addition to the cyanide-sensitive cytochrome chain also found in other organisms. Oxidation of NADH through this alternative pathway produces energy in the form of heat but no ATP. Some physiologists suggest that this pathway is a mechanism to prevent overreduction of the respiratory pathway, which would lead to the production of toxic free radicals. Others believe that this pathway allows the TCA cycle to continue at times of decreased need for ATP, to produce more than the usual amount of metabolites, which, in the presence of ATP, could not normally be produced. This system functions at a high rate in the flowers of a range of species, including the arum lily (Araceae). Temperatures of this organ may reach 40 °C (104 °F), which also contributes to the attraction of pollinators.

Photosynthesis

The autotrophic mode of nutrition of plants, is derived from oxygenic photosynthesis. Energy-rich organic compounds are synthesized from low-energy atmospheric CO_2, using the energy of absorbed sunlight. (Some bacteria are nonoxygenic photosynthesizers, utilizing hydrogen sulfide, H_2S, rather than water). The resultant organic compounds initiate the flow of energy and carbon through the food chains of managed and natural ecosystems, intrinsically linking plants with the heterotrophic life-forms of the remaining kingdoms of organisms. The oxygen liberated by plants (and certain photosynthetic protists and prokaryotes) over geologic time has oxygenated Earth's atmosphere and has produced fossil fuels such as coal, gas, and oil.

Basic Mechanisms

Electromagnetic radiation having wavelengths between approximately 400 and 700 nanometres can be seen as light by the eye and constitutes the range absorbed by plants for photosynthesis. Blue light has a wavelength around 450 nanometres, and red light, a wavelength of 650–700 nanometres.

Double-membraned cell organelles called chloroplasts contain the photosynthetic apparatus: Light-absorbing pigments, other electron-carrying chemicals (cytochromes and quinones), and enzymes. (Pigments absorb light of a particular wavelength; those wavelengths that are not absorbed are reflected and may be perceived as colour—hence, for example, the green colour of many plants). The inner membrane of the chloroplast is folded into flat tubes, the edges of which are joined to hollow sacklike disks called thylakoids. Stacks of thylakoids embedded with pigment molecules are called grana. The inner matrix of the chloroplast is called the stroma.

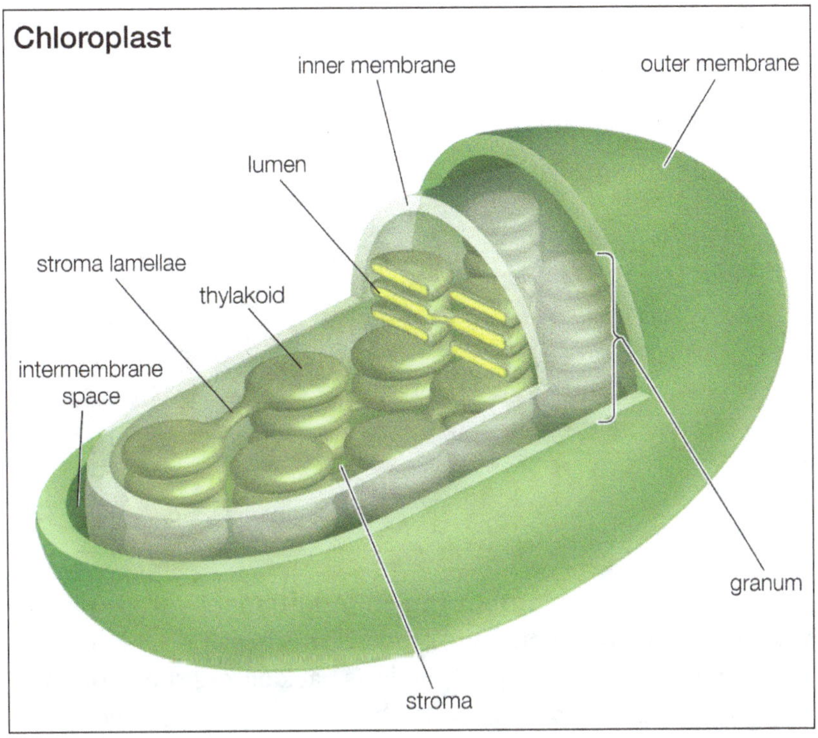

The internal (thylakoid) membrane vesicles are organized into stacks, which reside in a matrix known as the stroma. All the chlorophyll in the chloroplast is contained in the membranes of the thylakoid vesicles.

Photosynthesis consists of two interdependent series of reactions, the light, or light-harvesting, reactions and the dark, or carbon-assimilating, reactions; the former are dependent on light, the latter on temperature. Light reactions occur in the grana and dark reactions in the stroma. The overall formula for photosynthesis is:

$$6CO_2 + 12H_2O \rightarrow C_6H_{12}O_6 + 6O_2 + 6H_2O.$$

The light reactions, the first stage of photosynthesis, convert light energy into chemical energy (ATP and NADPH). Light reactions comprise two interdependent systems, called photosystems I and II. The dark reactions, the second stage of photosynthesis, use the chemical energy products of the light reactions to convert carbon from carbon dioxide to simple sugars.

The light reaction of photosynthesis. The light reaction occurs in two photosystems (units of chlorophyll molecules). Light energy (indicated by wavy arrows) absorbed by photosystem II causes the formation of high-energy electrons, which are transferred along a series of acceptor molecules in an electron transport chain to photosystem I. Photosystem II obtains replacement electrons from water molecules, resulting in their split into hydrogen ions (H+) and oxygen atoms. The oxygen atoms combine to form molecular oxygen (O_2), which is released into the atmosphere.

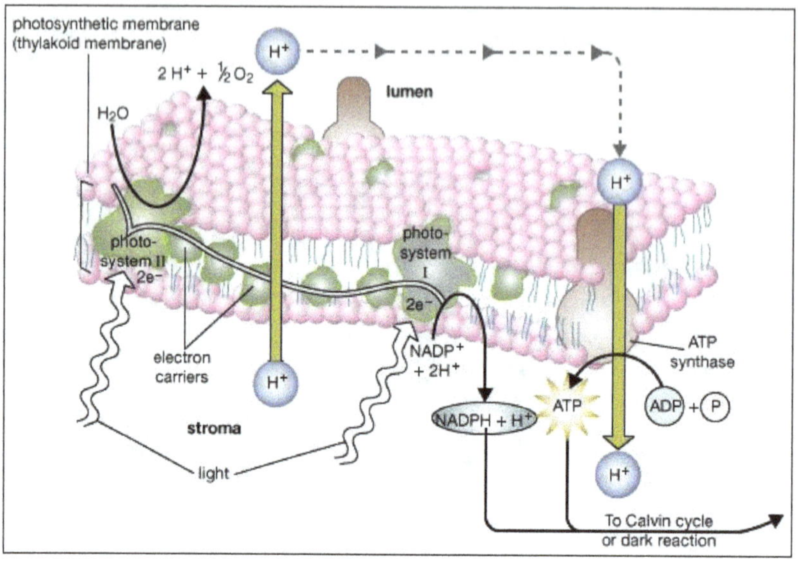

The hydrogen ions are released into the lumen. Additional hydrogen ions are pumped into the lumen by electron acceptor molecules. This creates a high concentration of ions inside the lumen. The flow of hydrogen ions back across the photosynthetic membrane provides the energy needed to drive the synthesis of the energy-rich molecule ATP. High-energy electrons, which are released as photosystem I absorbs light energy, are used to drive the synthesis of NADPH. Photosystem I obtains replacement electrons from the electron transport chain. ATP provides the energy and NADPH provides the hydrogen atoms needed to drive the subsequent photosynthetic dark reaction, or Calvin cycle.

Light reactions consist of several hundred light-absorbing pigment molecules arranged so as to maximize the gathering of light energy. These "antennae" are coupled to a mini-circuit of electron-carrying chemicals. The pigments are two types of chlorophyll—chlorophyll a and chlorophyll b—and various carotenoids. Absorbed light energy is transferred to specialized chlorophyll molecules called P_{700} and P_{680} in photosystems I and II, respectively. Once these specialized chlorophyll molecules have acquired sufficient energy, electrons are given up to the electron carriers within their photosystems, initiating an electron flow. (The carrier molecules include plastoquinones and cytochromes). The effect of this, when photosystems I and II function synchronously, is the formation of a chemiosmotic gradient of protons that phosphorylates (adds a phosphate group to) ADP, resulting in ATP. Those electrons also lead to the formation of NADPH from NADP. The P_{680} chlorophyll, upon loss of its electron, becomes a strong oxidizing agent that subsequently causes the water molecule to dissociate into protons and oxygen gas.

The dark reactions are responsible for the conversion of carbon dioxide to glucose. The essential reaction involves the combining of CO_2 with the five-carbon sugar ribulose 1, 5-bisphosphate (RuBP) in a series of reactions called the Calvin-Benson cycle. This reaction yields an unstable six-carbon intermediate, which immediately breaks down into two molecules of phosphoglycerate (PGA), a three-carbon acid. Each reaction is catalyzed by a specific enzyme. Six revolutions of the cycle means that 6 CO_2 molecules react with 6 RuBP molecules to produce 12 molecules of PGA; 2 three-carbon PGA molecules combine to form the six-carbon glucose, and 10 PGAs are recycled to regenerate 6 molecules of RuBP. The ATP and NADPH from light reactions provide the energy and reducing power to form glucose and refurbish the CO_2 acceptor, RuBP.

Specific Variations in Photosynthesis

Chlorophylls a and b (bound to proteins) and carotenoids constitute the principal light-absorbing complex of most plants. Differences in chloroplast structure, though not major, occur among phylogenetically diverse plant groups. All such variations, however, represent evolutionary adaptations to more efficiently utilize the light energy that drives the reactions common to all oxygenic photosynthesizers (i.e., photosystems I and II) or to avoid damage due to excessive light.

The enzyme ribulose 1, 5-bisphosphate carboxylase/oxygenase (Rubisco) catalyzes the formation of organic molecules from CO_2. As the major enzyme of all photosynthetic cells, Rubisco is the most abundant protein on Earth. There is, however, a major catalytic flaw in the ability of this enzyme to convert CO_2 to sugars. In the presence of molecular oxygen, Rubisco also catalyzes a reaction in which oxygen is introduced (i.e., it acts as an oxygenase), and CO_2 is formed rather than converted.

Rubisco evolved in photosynthetic organisms that lived in the atmosphere of primitive Earth, an atmosphere that contained only traces of molecular oxygen and plenty of carbon dioxide. As photosynthesis in the cyanobacteria of Precambrian times (about 4.6 billion years ago to 541 million years ago) oxygenated the atmosphere, the ratio of carbon dioxide to oxygen fell drastically, and Rubisco began to function more and more as an oxygenase. This greatly reduced the net fixation of CO_2 into sugars and, therefore, photosynthetic efficiency. Rubisco as an oxygenase splits RuBP into one PGA and a two-carbon acid phosphoglycolate, which initiates the photorespiratory carbon-oxidation cycle, or photorespiration. (This cycle probably evolved to recycle PGA back into the photosynthetic pathway, thereby preventing an even greater loss of carbon). Photorespiration involves three organelles (chloroplasts, peroxisomes, and mitochondria), each with unique transport mechanisms for the cycle's intermediates.

All plants are classified as C3 (plants that use only the Calvin-Benson cycle), C4 (plants that use an additional CO_2-fixation mechanism and the Calvin-Benson cycle), C3-C4 (plants intermediate between C3 and C4), and CAM (plants that have a nocturnal variant of the C4 pathway).

The majority of plants fix CO_2 directly into RuBP, and their first stable product is the three-carbon acid PGA—hence the designation C3. Those plants have an active photo-respiratory cycle, especially at high temperatures.

Sometime during the Oligocene Epoch (33.9 million to 23 million years ago), certain of the angiosperms (grasses and dicotyledonous plants) of mainly tropical climates evolved a CO_2-fixation system that acted ahead of the Calvin-Benson cycle. The first fixation is into the three-carbon acid phosphoenolpyruvate (PEP) by PEP carboxylase (an enzyme that has no oxygenase function) in the outer mesophyll cells of the leaf. The first stable fixation product is the four-carbon acid oxaloacetate—hence the designation C4 plants. Oxaloacetate is reduced to malate, which is transferred to a thick-walled bundle sheath cell. Malate is decarboxylated, giving rise to high CO_2 concentrations in the bundle sheath. Here, Rubisco of the Calvin-Benson cycle functions more efficiently because oxygenation is suppressed. There is thus a spatial separation of initial CO_2 fixation and the Calvin-Benson cycle. This efficiency is not without cost, however, as additional ATP is required to recycle PEP. For this reason, C3 plants may be more efficient in cold climates, where photorespiration is insignificant; under conditions where there is less available light, the higher ATP requirement would become a penalty.

The C4 pathway is effective at fixing CO_2 under drought conditions or under conditions where CO_2 is limited. C4 plants do not need to open their stomata as wide as C3 plants, because their primary carboxylating enzyme is saturated at much lower CO_2 concentrations. As a result, they lose less water during photosynthesis, and they are better able to cope in regions with arid climates. As humans continue to burn fossil fuels and thus increase the CO_2 concentration in the atmosphere, the relative advantage C4 plants enjoy in Earth's warm regions at present will diminish.

There are also plants with enzymatic and leaf anatomical characteristics intermediate between C3 and C4 plants, called C3-C4 intermediate species. Those plants are thought to be in the pathway of evolution to full C4 photosynthetic status.

Bishop's cap cactus (Astrophytum myriostigma).

Succulent plants of the desert regions (e.g., cacti) also initially fix CO_2 into oxaloacetate. This occurs only at night when conditions are cooler, however. Normally, the stomata in leaves or stems, through which plants lose water and acquire carbon dioxide, are open in the day and closed at night; however, the stomates of succulent plants that use the C4 pathway do the opposite and hence prevent loss of water during the hot days.

The resultant oxaloacetate is converted into malate, stored in the vacuole as malic acid, and released during the day when the stomates are closed. Malate is decarboxylated, and the CO_2 that is released is fixed by Rubisco in the usual Calvin-Benson cycle. Both the C4 and C3 processes take place in the same cell. This process is called crassulacean acid metabolism (hence CAM plants), after a family of succulent plants (Crassulaceae).

PRIMARY METABOLIC PATHWAYS

Carbohydrate Metabolism

Carbohydrates are defined as organic compounds consisting of polyhydroxy aldehydes or ketones or their derivatives. Carbohydrate means polymers of sugars plus water. The general formula for carbohydrates is $(CH_2O)n$. The most abundant organic compounds available in nature are carbohydrates and they are mostly produced by plants. Nearly 450 billion tons of carbon along with 13 x 1018 K.cals of solar energy is fixed into carbohydrates annually by land plants. Aquatic Plants (including marine) plants also fix substantial amount larger than land plants of dissolved carbon. Carbohydrates thus act as the primary food and fuel for all living organisms and also it is the main source of energy for various human needs like firewood, coal, biogas, petroleum, and bio-electricity etc. It also provides Carbon skeletal components for the other organic compounds.

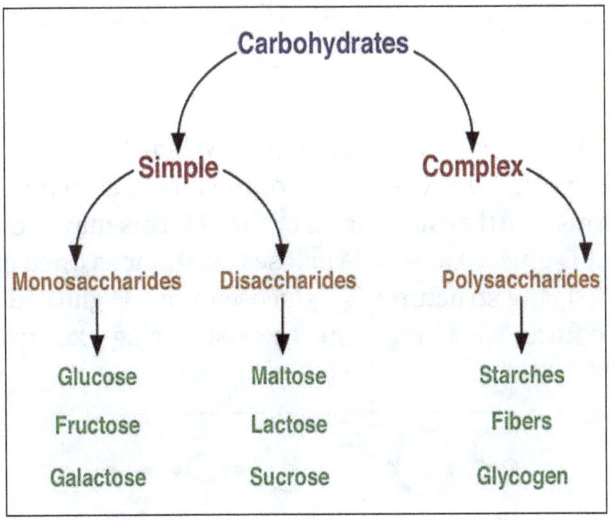

Basic Units of Carbohydrates

Basic units of carbohydrates are monosaccharides with an empirical formula of $(CH_2O)n$ where n=3 or more. Depending upon the number of carbon atoms present in a single molecule, they are named as Triose (3C), Tetrose (4C), Pentose (5C), Hexose (6C), Heptose (7C), etc. In the above said molecules, except one carbon atom all other carbons possess hydroxyl groups. If the carbonyl group (CHO) is present at the terminal region, such sugars are called Aldoses; on the contrary, if C=O group is present in any other position, they are called ketoses.

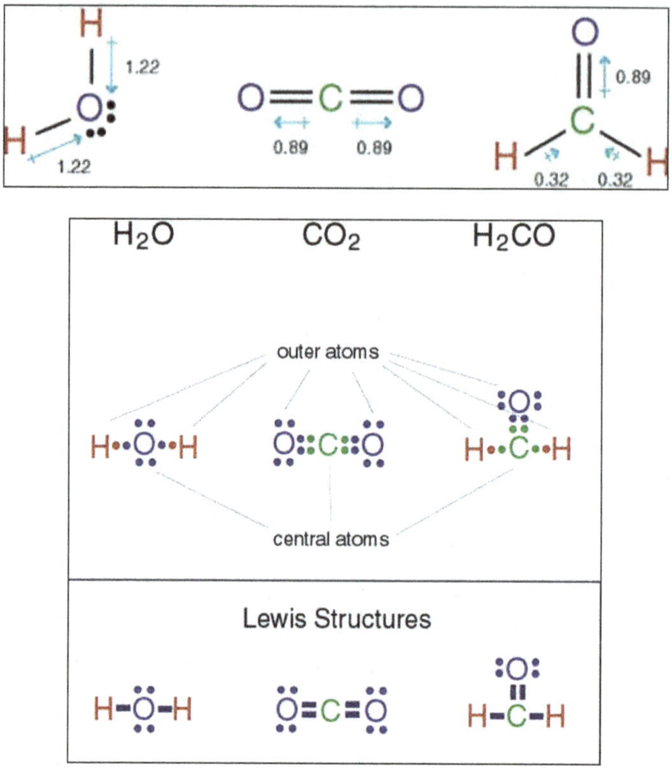

H_2O, CO_2 and H_2CO.

Almost all monosaccharaides exist as Aldoses as well as ketoses and each of them exhibit optical isomerism. Paradoxically, most of the monosaccharaides that are biologically active are D forms. Furthermore, each of the D forms may show A or B structures. Now it is known that D glucose and D-Fructoses exist not as open chain structures but as six members closed ring structures i.e. glucose as a – D glucose pyranose and fructose as a -D fructose furanose. Even pentoses exist as ring like structures. Eg. Ribose, Deoxyribose, etc.

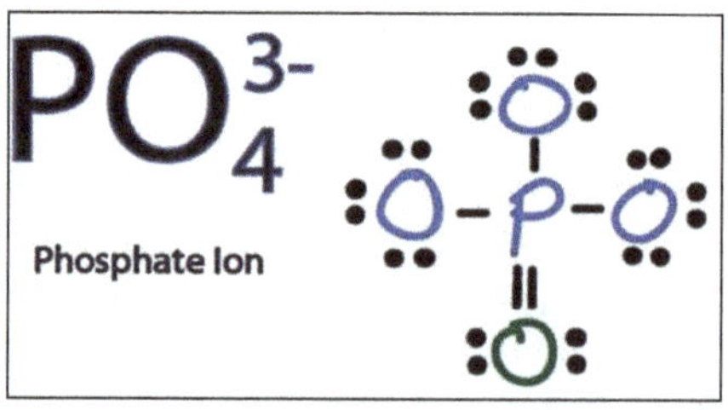

Gilber N. Lewis diagrams show bonding between atoms of a molecule and a lone pair of electrons if exist in a molecule Alewis structure can be drawn for any covalently linked moleucles and also to any compound with coordinate compounds.

Among the hexoses, glucoses act as building blocks for starch and cellulose, which form the bulk of the plant products. Among the pentoses, ribose and deoxyribose are the most important components of nucleic acid. Many monosaccharides also exist as glycosidic O-acyl or C-methyl derivatives and some may exist as alcohols, acids, phosphates or amino sugars.

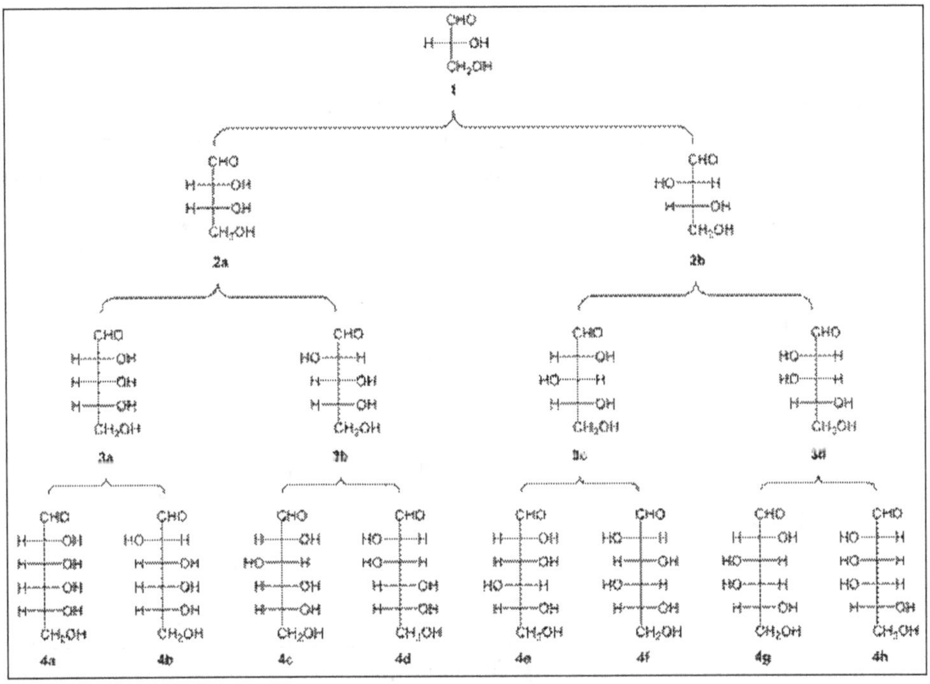

“Family tree” of the D -aldoses. By appending CH-OH groups to extend the basic framework, so more sugar can be derived (from trio with three C-hexoses to six carbon atoms) Here, the rotation of polarized light with (+) and) (specified -. (1) D -(+)- glyceraldehyde , (2) D -(-)- erythrose, (2b) D -(-)- threose, (3a) D -(-)- ribose , (3b) D - (-) - arabinose, (3c) D -(+)- xylose , (3d) D -(-)- lyxose , (4a) D -(+)- allose , (4b) D -(+)- altrose; (4c) D -(+)- glucose, (4d) D -(+)- mannose , (4e) D -(-)- gulose, (4f) D -(-)- idose; (4g) D - (+) - galactose, (4 h) D -(+)- talose; Basic monosaccharaides; Aldoses.

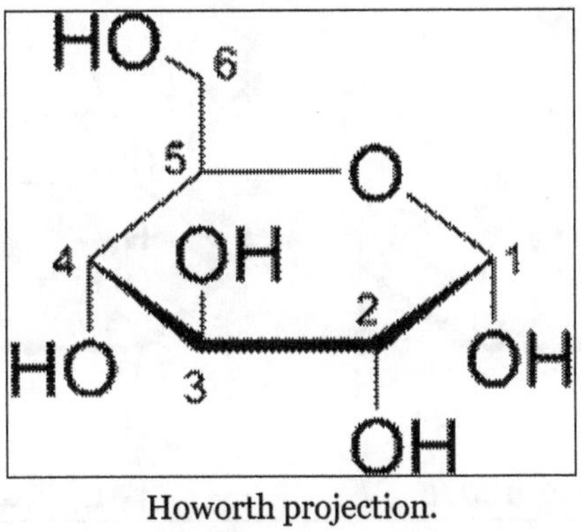

Howorth projection.

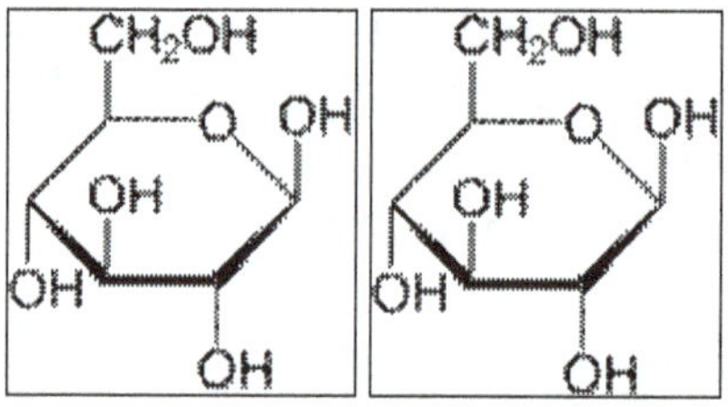

Alpha D Glucopyranose; Beta-D-glucopyranose.

The three-dimensional structure of monosaccharides in cyclic form is usually represented by its Haworth projection. In this diagram, the α-isomer has the OH- of the anomeric carbon below the plane of the carbon atoms, and the β-isomer has the OH- of the anomeric carbon above the plane. Pyranoses typically adopt a chair conformation, similar to cyclohexane. In this conformation the α-isomer has the OH- of the anomeric carbon in an axial position, whereas the β-isomer has the OH- of the anomeric carbon in equatorial position.

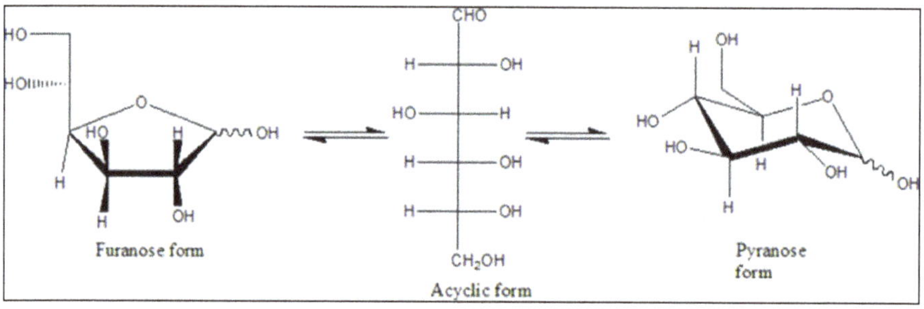

Furanose, Acyclic and Pyranose forms.

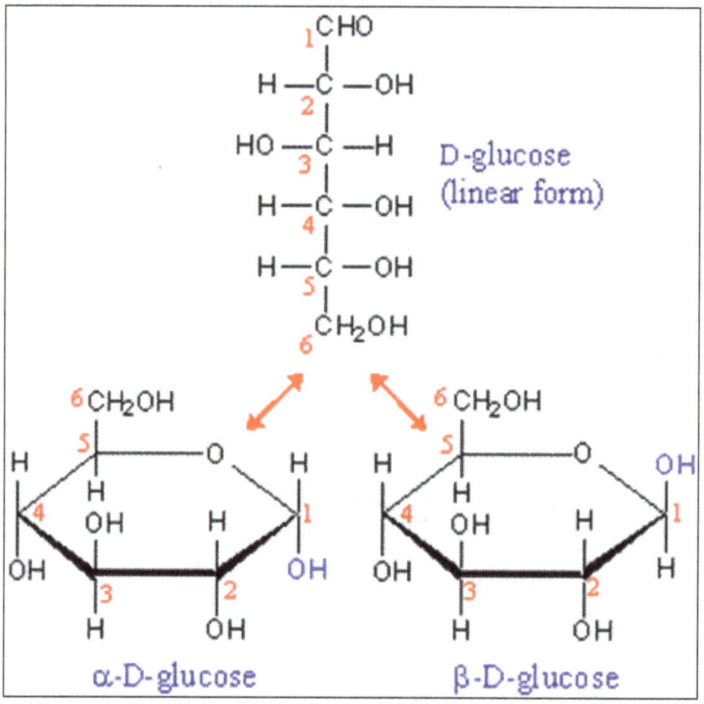

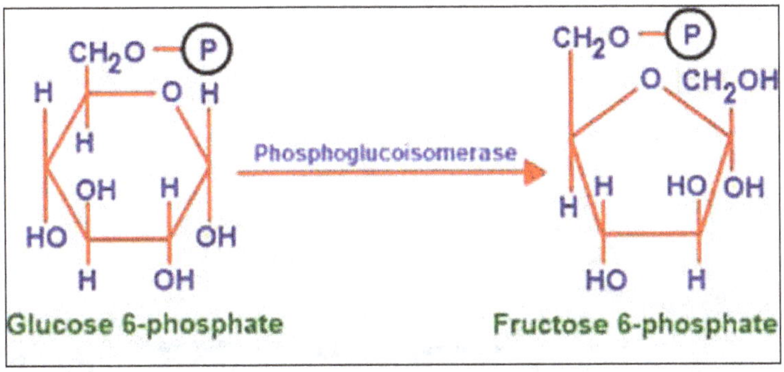

Di and Polysaccharides

Two monosaccharides, if linked by a glycosidic bond such a compound is called a disaccharide. If three are linked, then the compound is called a trisaccharide. The most common disaccharides found in plants are maltose, cellobioses sucrose and gentiobose. Among trisaccharide, raffinose and (melezitose?) are found in sugar beets, root tubers and coniferous trees.

Polysaccharides are those carbohydrates that are made up of three or more monosaccharides linked by glycosidic bonds. Plants contain a variety of polysaccharides of which starch and cellulose are the major carbohydrate components of the plant body.

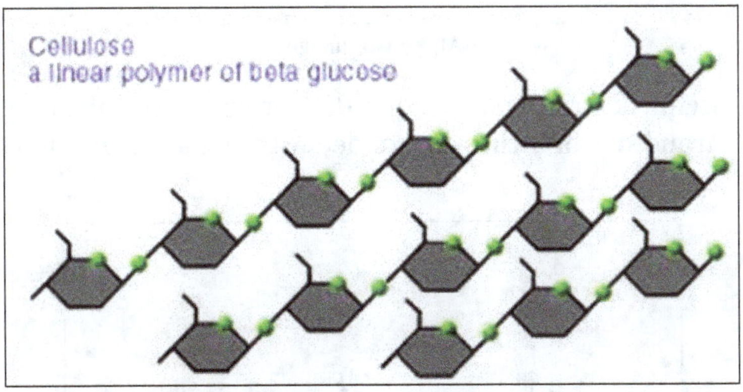

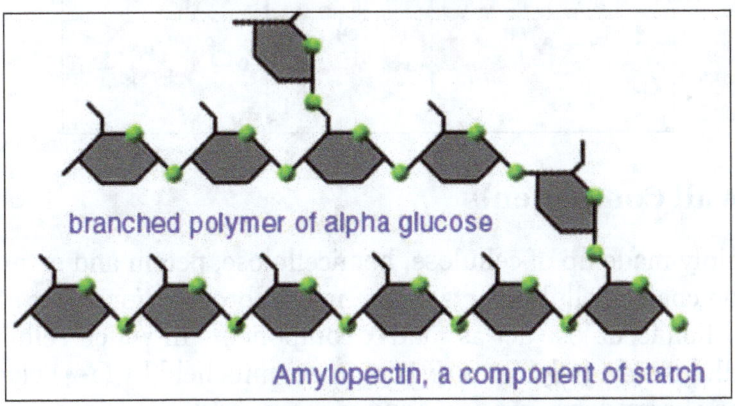

Oligosachharides

Raffinose and Raffinose Family

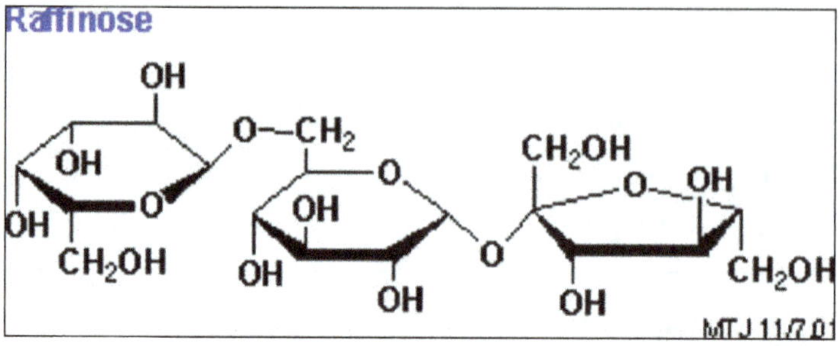

Raffinose.

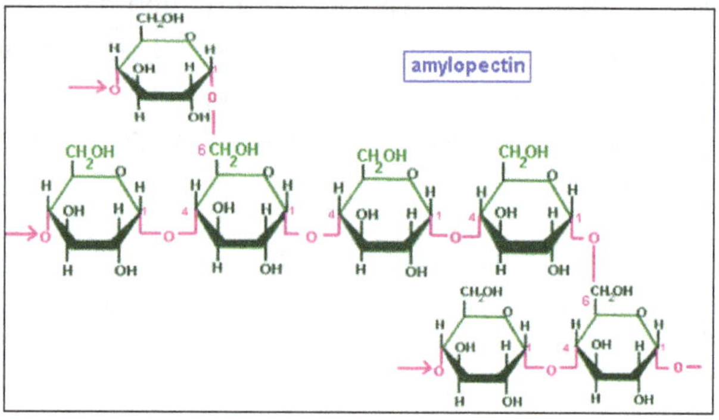

Alpha 1-6linkage.

Just like amylopectin, dextran a polysaccharide, also contains alpha 1-6 glycosidic linkage in main chain and side branches are bonded with alpha 1-3 and alpha 1-4 glycosidic linkage.

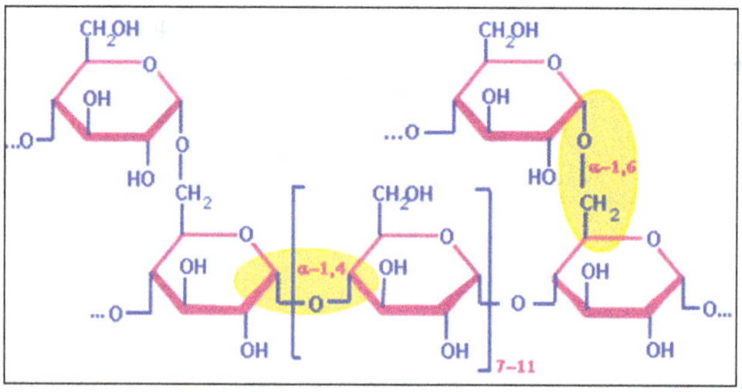

Plant Cell Wall Components

Cell wall is mainly made up of cellulose, hemicellulose, pectin and extensin, the latter is a polypeptide chain. Cellulose acts as the main fibrous component of the cell wall, but pectin and hemicellulose act as matrix components in which cellulose fibrils are embedded. Cellulose is a polymer of β-D glucose units held by (1-4) glycosidic bonds. Such chains may be made up of 1000-1500 glucose residues, whose mol. wt my range from 150,000 to 250,000 Daltons. Many such cellulose chains aggregate into bundles called micelles and micelles into micro fibrils. Micelles are interconnected with few cellulose fibers. Cellulose, though hydrophilic in nature, it is insoluble in water. Almost pure cellulose can be found in cotton fibers.

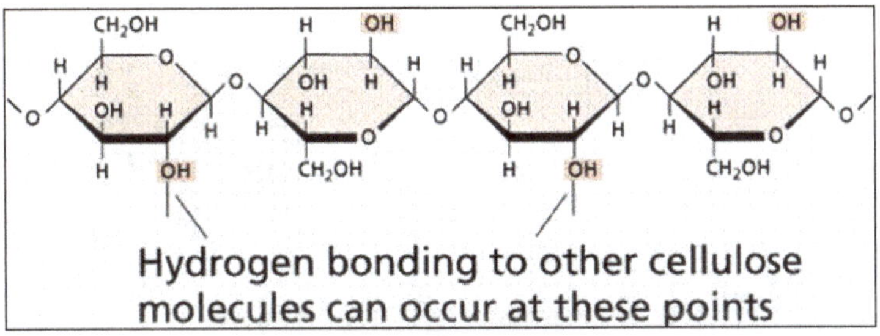

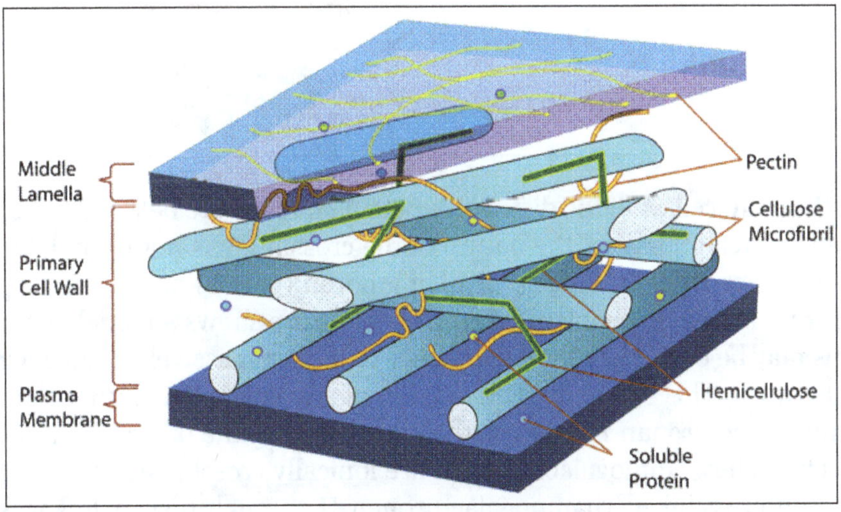

Pectin, like our common cment, is the major binding component of the cell walls of plants and fruits, It is chemically a polysaccharide, consisting of a linear chain of linked molecules of galacturonic acid; regions on the backbone with a lot of side chains are the so-called "hairy regions", and regions with little side chains are referred to as "smooth regions".

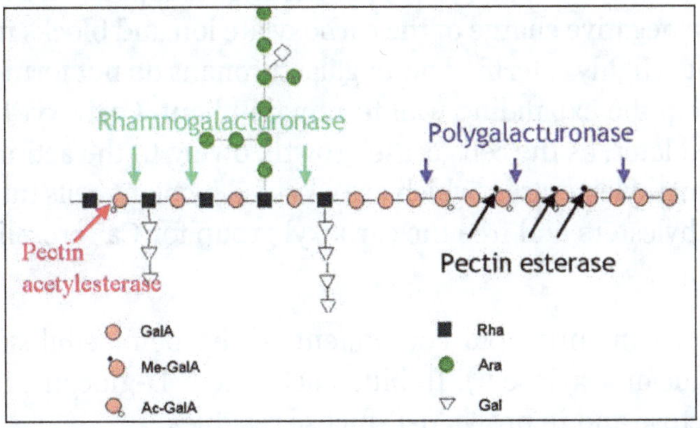

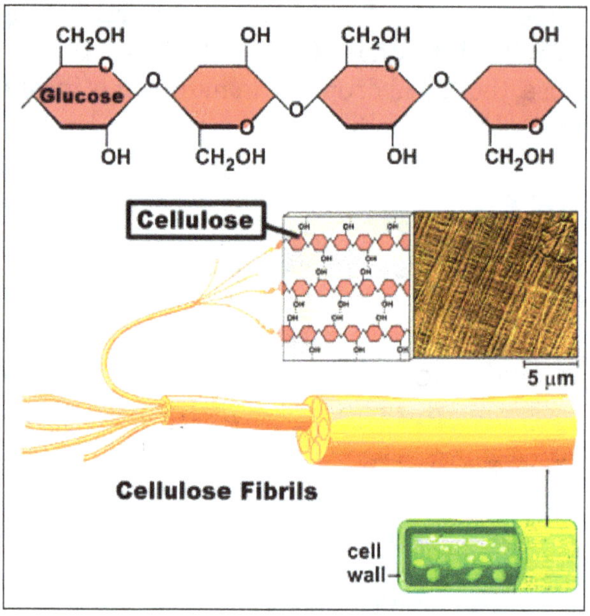

Growth of the plant cell wall; the distinctive pectin domains are believed to be covalently cross-linked to each other, but the nature of the crosslink has not been determined. In addition, two other types of linkage involving boron (part b) and calcium (part c) are important crosslinking mechanisms. Part a of the figure shows a model of how the pectin domains may be covalently linked together to form a massively large macromolecular pectin network. This is a simplified version of a recent model by Vincken et al.[4], in which rhamnogalacturonan I serves as the backbone and the other pectin domains are attached as branches. Homogalacturonans are ionically crosslinked by calcium (part c) whereas boron crosslinked rhamnogalacturonan II through diester linkages (part b). Rhamnogalacturonan II forms dimers through a borate ester bond (part b). This crosslinking is important for normal wall formation as well as for the control of wall porosity and wall thickness. Homogalacturonan (also known as polygalacturonic acid) forms stiff gels through Ca^{2+}-mediated crosslinking of its carboxyl groups through ionic and COORDINATE BONDS (part c). Growing cells usually synthesize homogalacturonan in which 75% of the carboxyl groups (COO^-) are methyl esterified (COOMe) — this modification removes the negative charge of the carboxylate ion and blocks its ability to undergo Ca^{2+}crosslinking. Highly esterified homogalacturonans do not form stiff gels and their secretion might help the expanding wall to remain pliant. Carboxyl-based crosslinking sites are unmasked later, as the cells cease growth, owing to the action of pectin methylesterases. These methylesterases, which are secreted by plant cells into their wall space, hydrolyse the methylesters and free the carboxyl group for Ca^{2+}crosslink formation and gel formation.

Structure of xyloglucan; principal component of the hemicelluloses. The heptamer block is shown (glucan 4-xylose 3). In blue backbone β-D-glucans; in red α-D-xylose; in black α-D-galactose and in brown α-L-fucose residues.

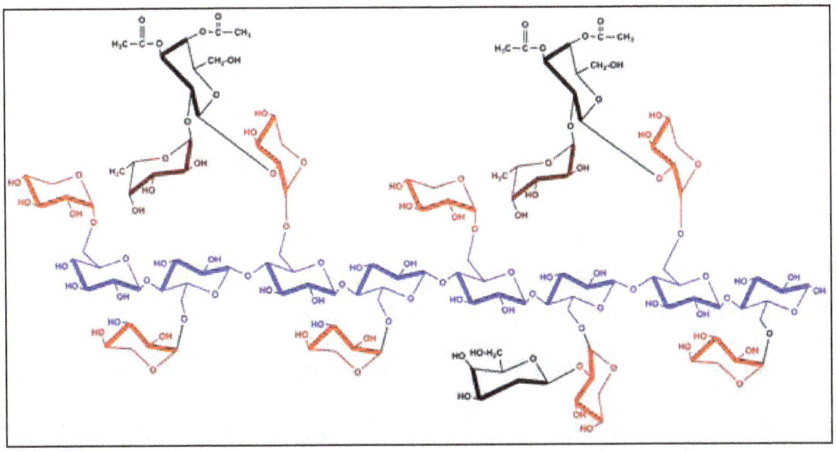

Lignin is a natural aromatic macromolecule in huge quantity and might serve as sustainable resources for the chemical industry after being depolymerized. An electrochemical approach combining anode oxidation and electro-generated H_2O_2 oxidation has been developed for converting lignin into value-added aromatic chemicals in this study. Lignin in alkali solution was electrolyzed in an undivided electrolytic cell with a cylindrical graphite felt cathode and a RuO_2–IrO_2/Ti mesh anode, in which the by-product O_2 on the anode could be efficiently reduced to H_2O_2 on the cathode in situ. Results display that the depolymerization productivity via the integrated approach obviously surpassed the sum of that by separate H_2O_2 oxidation and anode oxidation. Moreover, the analysis results of GC-MS, GPC, and C9 expanded formula confirmed that C–C bonds and C–O–C bonds in lignin were cleaved synergistically by direct anodic oxidation and indirect H_2O_2 oxidation, and the macromolecules are gradually depolymerized into final products of monomers and dimers.

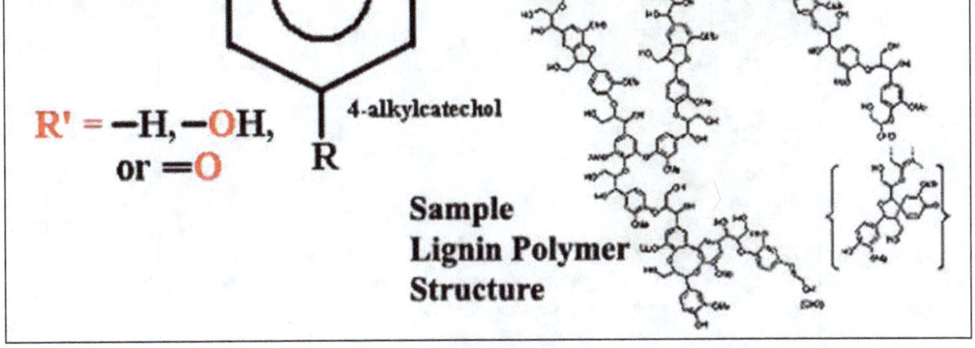

Lignin polymers.

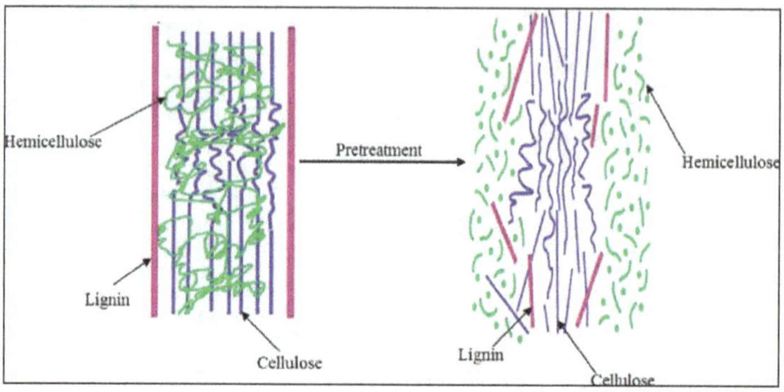

Glycerol treatment leads to breakdown products-Lignin, hemicellulose and cellulose.

This is a complex structure in which the cellulose is surrounded by a monolayer of hemicellulose and embedded in a matrix of hemicellulose and lignin. Furthermore lignin specifically creates a barrier to enzymatic attack while the highly crystalline structure of cellulose is insoluble in water while the hemicellulose and lignin create a protective sheath around the cellulose. This structure can be seen in the image below. This structure of lignocellulose therefore plays a huge role in inhibiting degradation of the hemicellulose and cellulose structure to monomeric sugars which is necessary to effectively convert biomass into ethanol. Processing of lignocellulose is therefore essential for the conversion of lignocellulosic biomass to biofuel such as bio-ethanol.

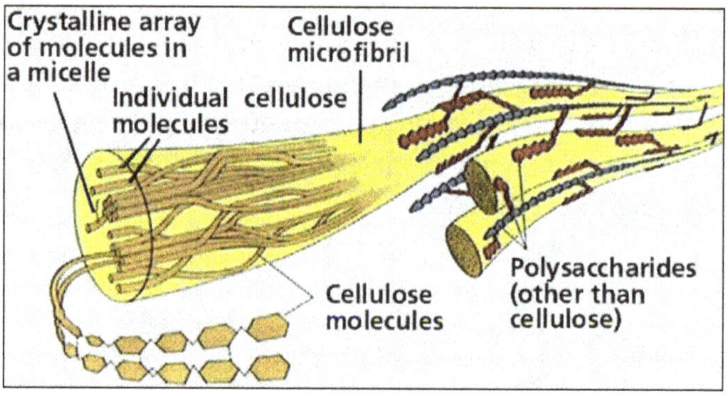

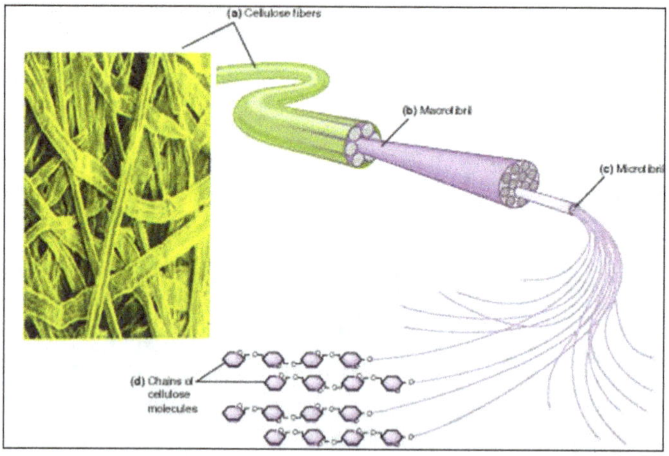

Cellulose structure is composed of a long, linear polymer of 1, 4 J3-linked glucose units that is found in the plant cell wall. Hydrogen bonding between sugar residues in adjacent parallel running cellulose chains imparts a micro fibril three-dimensional structure to cellulose. Being a large, linear, neutrally charged molecule, cellulose is water-insoluble, although it can be modified chemically (e.g., sodium carboxymethyl cellulose) to be more soluble and used as an additive in foods. Degradation of cellulose by colonic bacteria varies, but generally it is poorly fermented (doesn't digest well). Some examples of foods high in cellulose relative to other fibers include bran, legumes, peas, root vegetables, vegetables of the cabbage family, outer covering of seeds, and apples.

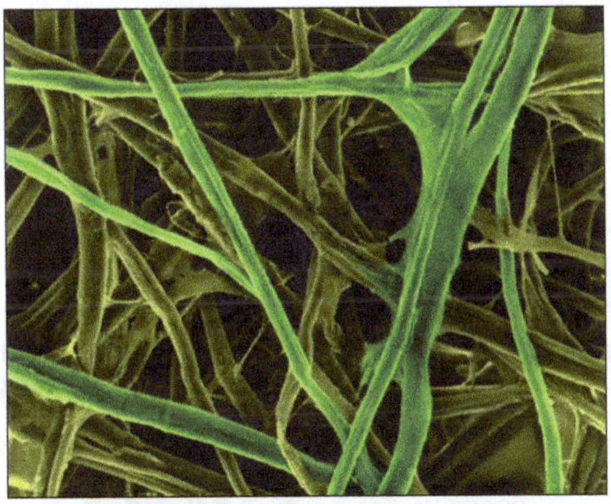

Cellulose fibers.

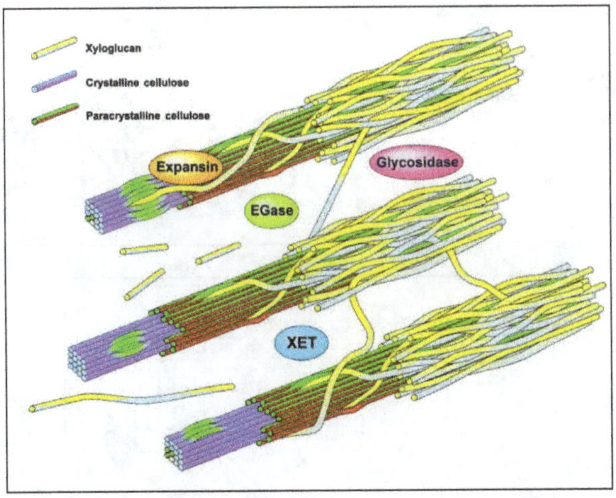

Modification of the plant primary cell wall is required for both cell expansion and for developmental events, such as fruit softening, where cell size remains static but where wall loosening is an important feature. Recent studies suggest that the cellulose–xyloglucan network is targeted by similar enzymatic activities in both expanding cells and ripening fruit but that unique isoforms are expressed in each process. Disassembly of this structural network probably involves the concerted and synergistic action of suites of these enzyme families, where one family of cell wall modifying proteins might mediate the activity of another, providing the basis for orchestrating ordered cell wall restructuring and turnover.

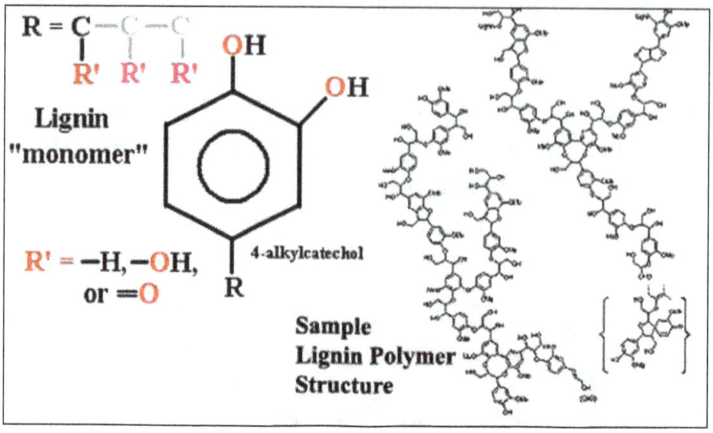

Lignin polymers.

Bamboo fiber, uses high quality natural bamboo as raw material, special tech process bamboo cellulose is extracted, then glue spinning processes and manufacturing regenerated cellulose fiber. Without any chemical additives, has good air permeability, water absorption, strong wear resistance, good dyeability features, known as the traditional four fibers (cotton, linen, wool, silk) five new fiber, breathing eco-textile, Fiber Queen, Bamboo fibers.

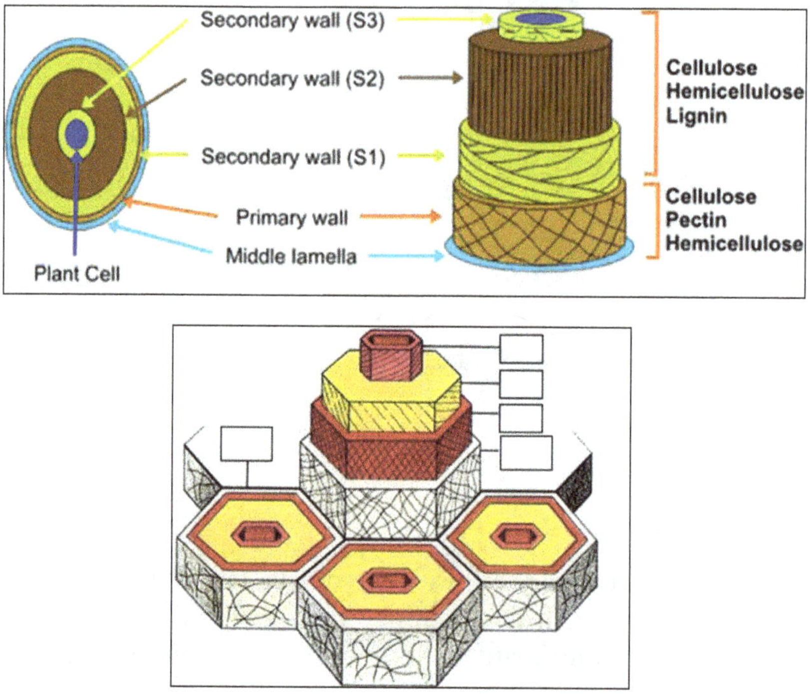

Cell wall layers-Primary to Secondary wall layers.

A group of cellulose fibrils in the form of micelles are surrounded by an amorphous matrix, which acts as the cementing material. It is mainly made up of polymers of rhamnogalactorunans, arabinogalactans, and polygalacturonans as pectic polymers and xyloglucans with arabinose and galactose as hemicellular materials. On the other hand, extensins is a hydroxyproline rich polypeptide chain, which act as cross link between cellulose and hemicellulose materials. Among cell wall polysaccharides, cellulose is extremely important for the mankind.

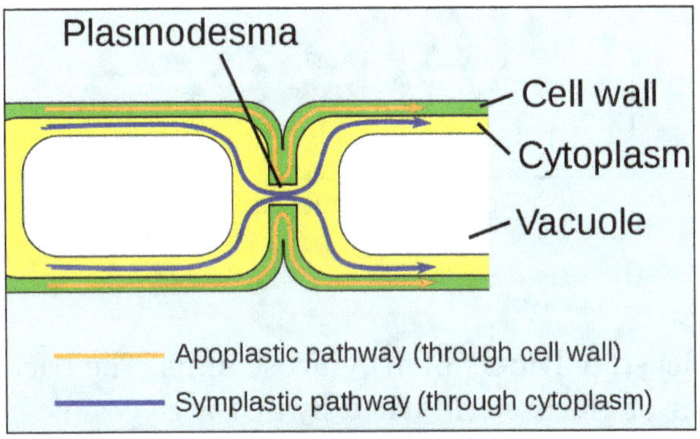

Plasmodesmata.

Recent studies have revealed that some oligosaccharides of the cell wall have regulatory role on cellular metabolic activities, growth and differentiation.

Starch

Plants synthesize and store starch as, reserve food material in specialized organs like root tubers, stem tubers, seeds and fruits. Starch granules are mostly deposited in amyloplasts. The important constituents of starch are amyloses and amylopectoses, the latter is always found in excess. The way nature of the pea seeds to be wrinkled and round is due to the presence of excess of starch, but the wrinkled is due to amylopectins.

Amylose

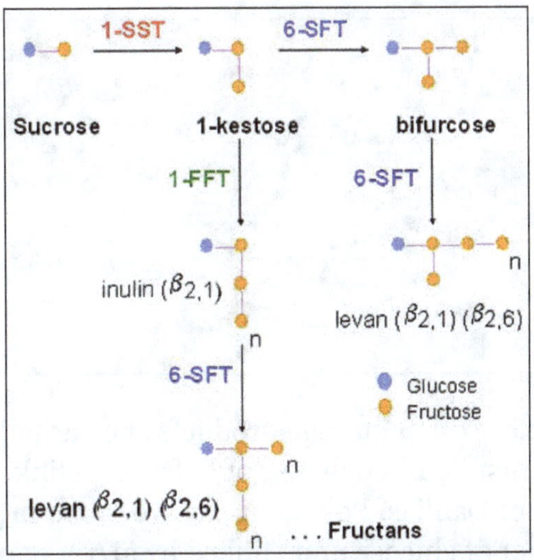

Amylose is an unbranched long chain of a -D glucose units held by a1 – 4 glycosidic bonds. These molecules show helical coiling with six glucose units per turn. They are not soluble in water but with water they form hydrated micelles, which give blue color with iodine.

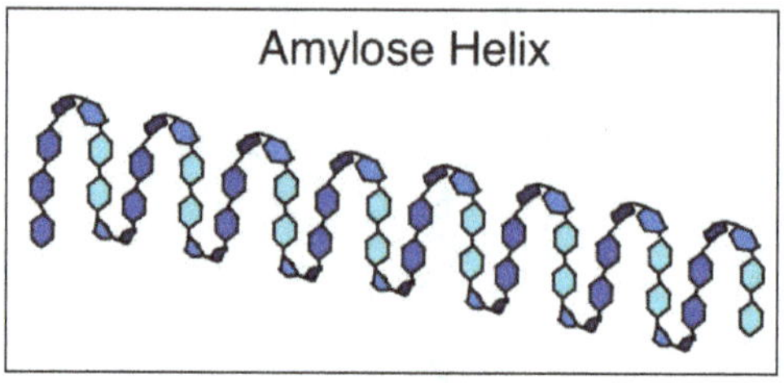

Eight glucose units per turn.

Amylopectose

They are highly branched polymers of aD-glucose units. The backbone of the chain is made up of a number of glucose units held by aD1-4 glycosidic bonds similar to that of amyloses. But at the interval of every 24-30 glucose residues a branch develops due to ab1-6 linkages with glucose units found on the backbone chain. Amylopectins are very large and often reach colloidal dimensions, when dissolved in water the solution becomes viscous. In the presence of iodine amylopectins give deep violet color.

There is not much difference between plant starch and animal starch, where animal starch is called glycogen, except for the fact that the branching in glycogen is extensive. The aD1->6 branching is at the interval of every 8-10 glucose residues. X-ray diffraction patterns of amylopectins show that polymer chains are helically coiled to each other, similar to that of double helix. In fact, such chains are deposited in amyloplasts in concentric layers, which starts from a common centre called hilum. In many lower algae, starch is deposited around a proteinaceous body called pyrenoid found within chloroplasts. The deposition of starch layers is species specific, which is determined by its genetic factors.

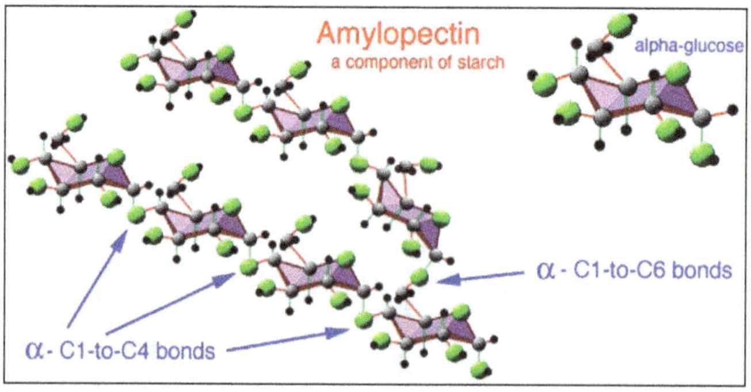

Apart from starch, there are other storage products such as polyfructans called levans, innulins, mannans, etc. Insulin is a polymer of D-fructose linked by β1->2 bonds. They are found in root tubers of Dahlias. Polymannans are found in yeasts, molds and some algae. Callose is a polymer of glucose units linked by aD1-3 bonds.

Inter Conversion of Carbohydrates

Synthesis and Degradation of Sucrose

Sucrose is one of the most important natural organic substances that are used daily by mankind as a sweetening agent, but sucrose as a nutrient, has its own caloric value. But in recent years, non-caloric sweeteners have been found; e.g. Saccharin and monellin. Saccharin is 400 times sweeter than sucrose, monellin extracted from an African serendipity berry is 2000 times sweeter than sucrose and monellin happens to be a protein (into 11,000). Its sweetness has been attributed to its 3-D conformation of its polypeptide chain. When monellin is heated to denaturation, it looses its sweetness. Such non caloric sweeteners are very useful for diabetic patients. However, sucrose is a natural compound synthesized and stored in various organs of different plants, like sugarcane, sugar beet, sweet potato, etc. In fact plants, transport their photosynthate from the region of synthesis to the region of need in the form of sucrose or sucrose phosphates.

Though glucose and fructose are the components of sucrose, the synthesis of sucrose requires uridine diphosphate a-D glucose (UDPG) and β-fructose 6-phosphate. The condensation reaction is performed by sucrose phosphate synthetase. The sucrose 6-phosphate is then dephosphorylatated by sucrose 6-phosphotase enzyme.

However, the major pathway that operates in many plants is actually the reverse of sucrose breakdown. This method involves the condensation of UDPG and fructose directly. The enzyme involved in this reaction in sucrose synthetase. On the other hand, sucrose is hydrolyzed by sucrose phosphorylase or invertase.

Synthesis of Starch

Plants synthesize starch as the major reserve food product in seeds and tubers. In lower plants, the synthesis and accumulation of starch takes place around pyrenoids found in chloroplasts. In higher plants though there is no such localization, starch is generally synthesized in chloroplasts or in amyloplasts, but Amyloplasts store starch as the major stored polysaccharide. The mechanism of starch synthesis has been studied extensively in the fruits of Zea mays and Oryza sativa and in the leaves of spinach, etc. In the above said plants it has been demonstrated that the starch synthesis is medicated by an enzyme called starch syntheses which exists in two isozyme forms. One utilizes ADP-Glucose to start with and the other uses Uridine-Di Phospho Glucose (UDPG). Though they require a primer (a short chain of glycans) for their activities, they can still perform polymerization of glucoses residues without them for they can prime such reactions.

Under normal cellular conditions the enzyme adds ADPG or UDPG to aD glucose present at the non reducing end of the primer and forms a1-4 glycosidic bonds. By this method, the enzyme goes on adding glucose units one after another, which results in the formation of an unbranched chain of glucose units called amylose. The synthesis of amylopectins also takes place along with the synthesis of amylose. The enzyme responsible for the branch formation is called Q-enzyme which is similar to that of branching enzyme found in animal systems. The Q-enzyme, isolated from starch containing tubers, has the ability to form a 1-4 as well as a 1-6 glycosidic bonds, simultaneously because both the enzyme forms are tightly bound to each other. However, recent studies on the nature of waxy and sweet corn mutants in Zea mays show that the waxy variety has greater amount of amylopectoses. The above observations suggest that there are to different enzymes for the synthesis of Amylose and Amylopectin and their activity is also independent of each other.

Plants also possess another pathway where amylose chains are synthesized by the activity of starch phosphorylase enzyme, which uses glucose 1, P and the same are added to the non reducing end of the primer chain. Thus the amylose chain grows. Cory and Cory have done extensive studies on these enzymes in animal systems. It is now almost accepted that the starch phosphorylase enzyme found in plants is similar to that of animal enzymes with minor variations; the activity of this enzyme is regulated by the levels of inorganic phosphates and some phytohormones.

Degradation of Starch

In plant cells starch is degraded by the activities of a amylase, β amylase and starch phosphorylase enzymes. The activities of these enzymes may be independent or cumulative. Germinating seeds show very high activity of the above mentioned enzymes, but such enzymes are also found in the mesophyll cells of leaves.

The activity of starch phosphorylase enzyme is observed in both animal and plant cells; where amylose chains are hydrolyzed from the non reducing ends. The remarkable feature of these enzymes is that it utilizes inorganic phosphates and the same is added to glucose to yield glucose phosphate units. Such enzymes are also very active in guard cells during the opening of stomata. However, starch phosphorylase is incapable of degrading branching units.

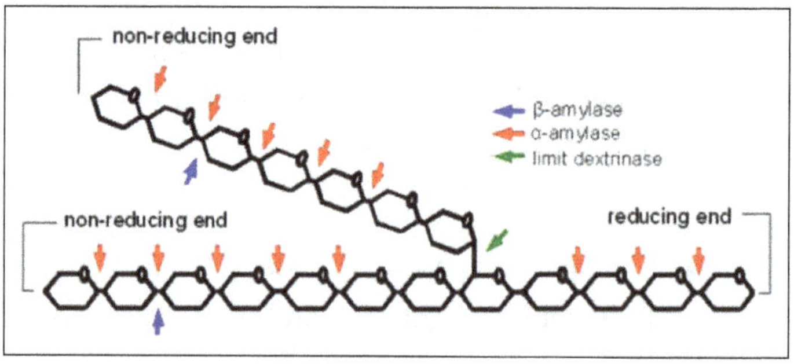

A large branched dextrin that shows which links can be broken by which of the mash enzymes.

Hydrolysis of starch by the activity of amylase enzymes is very interesting. While a amylase enzyme hydrolyzes chains randomly releasing single glucose units or maltoses, β-amylases hydrolyze amylose chains from non–reducing ends and the products are always maltose units. These enzymes are incapable of breaking the α1-6 branch chains. However, there are specific debranching enzymes called R-enzymes. They break a 1-6 bonds and thus complete the breakdown of starch. Later maltoses are also degraded into glucoses by the activity of maltase enzymes.

Germination of cereal grains results in the synthesis of a-amylase enzyme denovo. The synthesis of these enzymes is under the control of Gibberillic acid induced gene activation. The GA induced synthesis of a-amylase can be inhibited by the inhibitors like actinomycin D and cycloheximide, which inhibit transcription and translation respectively. The site of synthesis of the above enzyme is aleurone cells. The GA Produced by embryos diffuses into aleurone cells, where specific genes are activated. The resultant products like a amylases are secreted into endospermous tissue where the starch gets degraded into glucose and the same is used up by the growing embryo. The endosperm is a dead tissue. This work has generated a lot of interest in understanding how other hydrolyzing enzyme activities are regulated during germination and the development of plant body.

Inulin is another reserve food material found in the tubers of artichoke, dahlia, chicory and other composites members. It is made up of an unbranched chain of 35 fructose units linked by β 2-1 glycosidic bonds. Interestingly, the Inulin chain incorporates one molecule of glucose in the middle or at one end of the chain. The enzymes responsible for the synthesis and degradation of Inulin have been detected in many tubers containing Inulin as the reserve food material.

Synthesis and Degradation of Cell Wall Compounds

Cellulose is an important component of the cell walls. It is a polymer of β-D glucose units linked by β 1-4 glycosidic bonds. Cellulose and other cell wall carbohydrates form the bulk of the human diet. Unfortunately human beings are incapable of digesting cellulose and other related cell wall materials. Thus, in terms of calorie value, cellulose is useless as human diet. But for grass grazing animals it is the staple food, for they have the ability digest the cellulose and hemicelluloses in the intestine by the colon bacteria they have. Can you imagine, if human beings are introduced with such harmless bacteria, we can live on grass; world population does not and need not suffer from food shortage.

Synthesis

The synthesis of cellulose has been studied under in vivo and invitro conditions in many systems such as mung bean, lupinus, pea plants and also in some bacteria. Though βD-glucoses are the main precursors, even mannose, galactose and other hexoses are utilized in the cell wall formation by converting them into glucoses. The enzyme called βD-1-4 glyconyl transferase which is also called cellulose synthase, is responsible for the synthesis of cellulose. The enzyme utilizes GDP glucose or UDP glucose as the free units, and the same are linked to cellobiose units by the formation of βD1-4 glycosidic bonds. Here cellobiose units act as primers. Thus a long chain of cellulose fibers develop. The cellulose synthesizing enzymes are located in the membrane fractions of the plant cells. In fact, Golgi vesicles get loaded with enzymes and the precursor hexoses, and then they are transported to the plasma membrane where the vesicles by exocytosis secrete the enzymes and cell wall components. The transportation of Golgi vesicles is greatly aided by microtubules found in the cytoplasm. In recent years attempts to synthesize pure cellulose fibrils in test tubes have met with great success. Perhaps in another decade or so, cellulose fibers needed for cloth making will be artificially produced in industries with the use of sophisticated bio-techniques.

Degradation

Cellulose is the primary component of primary wall, but secondary walls are made up of cellulose, hemicellulose, lignin and other components. However, cellulose is degraded by the activity of cellulase enzyme which exists in two forms. One is active at acidic pH i.e. 4.5 to 5.0 and the other form is active at pH 5.9 to 6.3. They are located at the outer surface of the plasma membranes. These enzymes hydrolyze the B 1.-4 glycosidic bonds randomly and produce disaccharides or small oligosaccharides called cellobiose. Then cellobiase enzymes break down collobioses into glucose units.

The synthesis and activity of cellulase enzymes, however is under the control of phytohormones like auxins. During auxin induced cell elongation the acidic form of cellulases becomes active and the cell wall is loosened by hydrolytic activity. In addition, auxin induced cellulase synthesis by specific gene activation. Similarly, ethylene is another hormone which induces the synthesis of cellulase and pectinase denovo during fruit ripening. Even Abscisic acid induces cellulase synthesis during abscission layer formation in the stalks of leaves and fruits. The synthesis and degradation of other cell wall components like Polymannans, polyxylans, polyarabons, lignin and other pectic components have been observed and they are found to be under gene regulation.

Central Role of Glucose

Glucose plays a central role in the inter conversions of simple monosaccharides into various polysaccharides and vice versa. Interconversions are highly regulated. This depends upon the metabolic state or the needs of cells. High energy nucleotides like ATP, GTP and UTP are absolutely necessary in the synthetic process of carbohydrates. Further more glucose is oxidized to produce energy rich compounds. Sometimes many intermediary carbon compounds, required for the synthesis of many amino acids and other organic compounds are also produced. Under certain conditions, amino acids and other components are utilized in the synthesis of glucoses by the reverse pathway of glycolysis to produce glucose, which is known to be the ultimate product of photosynthesis, where solar energy is stored in the form of chemical energy (in chemical bonds) not only provides chemical energy but also acts as the structural components for many cell structures.

Polyamine Catabolism

Polyamines (PAs) are aliphatic amines found in all living cells. The most abundant PAs in plants are putrescine (PUT), spermidine (SPD) and spermine (SPM), the latter two are also called as higher PAs. However, the levels of individual PAs vary depending on the plant species or organ and on the developmental stage. The biosynthetic pathway of these PAs is well documented. Briefly, PUT is synthesized by the decarboxylation of ornithine, catalysed by ornithine decarboxylase (ODC), or indirectly by the decarboxylation of arginine by arginine decarboxylase (ADC), via agmatine. Higher PAs (SPD and SPM) are produced by the sequential addition of aminopropyl moieties to the PUT skeleton through enzymatic reactions catalyzed by the SPD and SPM synthases. S-adenosylmethionine decarboxylase (SAMDC) is responsible for the synthesis of decarboxylated S-adenosylmethionine which is used for the addition of the aminopropyl moiety. Besides the diamine oxidases (DAOs) and PA oxidase (PAOs), which catalyse the terminal catabolism of PAs in the apoplast, PAs also undergo rapid interconvertion in the PA-cycle in the peroxisomes, where the partial and/or full back-conversion of SPM to SPD and of SPD to PUT are catalyse by special PAOs. The metabolism of PAs in fungi shows some differences compared to plants. Compare to plants, not all fungi contain SPM. In fungi SPD is the most important PA, which involved in all the functions such as the regulation of cell proliferation and differentiation, vegetative growth, spore germination, sporulation or dimorphic transition, probably in all fungi. Another pronounced difference is, that in plants PUT is synthesized by two alternative pathways catalysed by ODC and ADC, but in fungi, there is a unique pathway PUT formation in which the enzyme ODC plays the central role. The synthesis of higher PAs (SPD and SPM) functions similarly as in plants, as also the terminal catabolism of PAs. Although completed with some additional enzymatic reactions the back-conversion of SPM into SPD, and SPD to PUT is also occurred in fungi.

Due to their cationic nature, they are able to interact with negatively charged macromolecules (such as nucleic acids, phospholipids, proteins) and stabilize their structure, especially under stress conditions. According to these bindings, PAs influence the membrane permeability characteristics, induce changes in protein function. Besides their direct protective role, their binding to the chromatin cause alteration in the availability of genomic sites to DNA or RNA polymerases; furthermore several protein kinases have been shown to be influenced at the transcriptional or post-translational level by PAs. In addition PA metabolism is also related to several other metabolic pathways, such as proline, and the synthesis of plant hormones and signalling molecules, such as ethylene, H_2O_2 or NO. Although several studies have been published on both the changes of PA levels during abiotic stress and protective effects of PAs against various abiotic stress factors, such as drought or cadmium, the mechanism by which PAs act during plant-microbe interaction is sometimes contradictory. In addition, the several types of the interactions between the plants and the microbes resulted in a divers and complex action mechanisms.

PAs are not only essential to maintain cell viability, but function as signalling molecules regulating many of the responses that help the plants to cope with biotic stresses and also to build symbiotic interactions. The present review focuses on this aspect of the mode of action of PAs and attempts to find answers to the following questions: Is there a relationship between accumulation of PAs and plant biotic stress resistance and what are the differences between biotroph and necrotroph interactions?

Plant-pathogen Interactions

Biotrophic pathogens: Biotrophic fungal pathogens colonize and require living plant cells to survive because they obtain nutrients from living host cells. During a compatible interaction between plants and pathogens, in the green island regions - surrounding the infection sites, induction of photosynthesis and delay in senescence were found - fungal pathogens induces the PA synthesis of the host, as it was found in the case of the compatible interaction between barley and Puccinia hordei Otth. or Blumeria graminis f. sp. hordei. Similarly, accumulation of the free and conjugated forms of SPD was found already after 3 days and of SPM 7 days after powdery mildew infection, in both susceptible and resistant wheat genotypes. As the precursor of SPD and SPM, PUT is, probably used up for synthesis of SPD and SPM, which were then stored in conjugated form. This suggested that the accumulation of PAs was a general response to inoculation; furthermore, differences between the initial and induced levels of PAs were not correlated with the degree of tolerance to powdery mildew. However, the opposite was found, when powdery mildew resistant and susceptible barley cultivars have been described, and the resistant cultivar accumulates higher PUT and SPD levels than the susceptible one after infection. Relationships between the PA contents and the level of pathogen resistance were also found in oat cultivars. Both resistant and susceptible oat cultivars in response to Blumeria graminis f. sp. avenae or Blumeria graminis f.sp. hordey, showed significant differences in the free PA levels. The resistant genotype showed higher level of SPD than the sensitive one, and showed a significant increase in it after Blumeria graminis f.sp. avenae (host interaction) inoculation, and similar pattern was detected for ADC activity. In the case of Blumeria graminis f. sp. hordey (non-host interaction) infection increased levels of PUT was found in the sensitive genotype. Increased levels of free PUT and SPD and of conjugated forms of PUT, SPD and SPM have already been reported 1-4 d after infection in an incompatible interaction, between powdery mildew and barley. In addition not only increased level of PAs, but increased activity of the synthetic enzymes, ODC, ADC and SAMDC were detected. The early accumulation of PUT and induction of ODC and ADC gene expression were found in wheat after Fusarium graminearum inoculation.

Changes in PA levels have also been found in oat after crown rust inoculation, and differences in PA contents between resistant and susceptible cultivars were associated with increased pre-penetration and penetration resistance. Increased amount of SPD and SPM, together with increased level of PA degradation were found in the resistant cultivar at early stages of the infection process. In addition, increased ADC activity in the susceptible cultivar was detected during pre-penetration, and exogenous PA treatment reduced the appressorium formation in the same oat cultivars. In most cases induction of PA synthesis has been reported during biotroph pathogen infection, however, in the highly susceptible sugarcane infected with Ustilago scitaminea reduced free PA content has been reported.

Increased gene expression level of genes encoding PA synthesis enzymes was found in flax seedlings infected with pathogenic strains of Fusarium culmorum and Fusarium oxysporum and the non-pathogenic strain of Fusarium oxysporum. The highest induction of the mRNA level was found in the case of ODC and ADC after Fusarium culmorum inoculation, while the gene expression of PAO did not change. However, the highest total PA level, especially cell wall-bound SPD content was found in the plants infected by the pathogenic Fusarium oxysporum, suggesting that locally accumulated PAs can directly inhibit pathogen growth or reinforce the cell wall to restrict pathogen invasion. Lowest induction in the gene expression levels and PA accumulation was detected in the case of the non-pathogenic Fusarium oxysporum strain. Differences observed between the two pathogenic strains also suggested, that the two pathogens have different mechanisms for plant tissues penetration, and they also induced different defence mechanism in flax seedlings.

According to these, fungal infection of plants might be controlled by a specific inhibitor of PA synthesis. ODC (responsible for PUT synthesis in fungi) enzyme can be irreversibly blocked by alpha-difluoromethylornithine, which has no effect on ADC DFMO provided protection against three biotrophic fungal pathogens, namely leaf rust, stem rust and powdery mildew in wheat. Originally, it was thought that inhibition of PA synthesis alone would be enough to control the growing of pathogen fungi. However, these treatments were not always completely efficient against pathogens, which may be due to an increase in the PA uptake by the pathogens from host. When wheat spikes were treated with DFMO and subsequently inoculated with Fusarium graminearum conidia, DFMO did not display inhibitory effects on the fungus. Indeed DFMO added to the medium induced the expression of a transcriptional regulator and the trichodiene synthase, which have role in deoxynivalenol mycotoxin biosynthesis. In addition up-regulation of ODC, and some type of PA transporter was found in F. graminearum cultures when treated with DFMO. These results suggested that the inhibition of PA biosynthesis together with inhibition of plant PA transporters, may be useful target as a way to avoid this secondary effect.

Similarly to the fungal pathogens, biotrophic interactions between viruses or bacteria and plants also revealed that during HR and the induced programmed cell death (PCD) the activity of ODC and the level of PAs increased in tobacco plants infected with tobacco mosaic virus, or in pepper plants after infection with Xanthomonas campestris. In tobacco during infection with the biotrophic bacterium Pseudomonas viridiflava an increase of apoplastic SPM levels was detected. Enhancement of host PA levels either due to overexpression of ADC gene or PAs treatments, decreased the growth of bacteria, while inhibition PAO enzymes enhanced it.

As the generation of H_2O_2 via the terminal catabolism of PAs plays a role in PA-related signalling processes, PA metabolisms may be involved in penetration resistance against biotroph pathogens. The formation of H_2O_2 resulting from the enhanced activities of DAO and PAO, could be responsible for the hypersensitive reaction (HR) observed in barley after powdery mildew infection. It has also been suggested that SPM is the most important PA in plant defence against pathogens in incompatible host-pathogen interactions. Therefore, it is an interesting aspect to examine the role of different PAOs in controlling HR and/or PCD in plants during pathogen attacks.

Necrotroph pathogens: Necrotroph pathogens kill the host cells at very early stages in the infection and feed on the remains. According to this it seems that PCD in the host would merely make life easier for the pathogen. Infection of tobacco by the Sclerotinia sclerotiorum resulted in increased gene expression level and activity of ADC in the host tissues, as well as PUT and SPM accumulation in the apoplast. The increased PA exodus and the apoplastic terminal catabolism has also role in the pathogen-induced necrosis. It was also found that overexpression of ADC in transgenic tobacco plants or infiltration with exogenous PAs, led to increased necrosis after infection with S. sclerotiorum. In contrast, inhibition of apoplastic DAO and PAO decreased leaf necrosis during the fungal infection.

However, the results on necrotroph pathogens are sometimes contradictory. Infection of tobacco leaves with Peronospora tabacina, Alternaria tenuis, Erysiphe cichoracearum, Pseudomonas tabaci or tobacco mosaic virus resulted in lower level of PUT and SPD in the infected plants, compared to the control. These results suggested that the changes were not related to the type of pathogens (fungi, bacterium or virus) or the type of mechanism (biotroph or necrotroph) but to the severity of damage symptoms. The SPD synthase over expressor transgenic tomato lines, which accumulate higher levels of SPD than the wild-type plants were more susceptible to Botrytis cinerea (necrotroph), while responses to Alternaria solani (necrotroph) or Pseudomonas syringae pv tomato (biotroph) were similar to the wild-type plants The inhibition of DAO activity in chickpea resistant to the necrotroph fungal pathogen, Ascochyta rabiei resulted in reduced resistance to the pathogen. In accordance with this, constitutive overexpression of cotton PAO gene- responsible for the back-conversion of SPM to SPD - in Arabidopsis improved resistance to Verticillium dahlia.

Polyamines (PAs) are aliphatic amines found in all living cells. The most abundant PAs in plants are putrescine (PUT), spermidine (SPD) and spermine (SPM), the latter two are also called as higher PAs. However, the levels of individual PAs vary depending on the plant species or organ and on the developmental stage. The biosynthetic pathway of these PAs is well documented. Briefly, PUT is synthesized by the decarboxylation of ornithine, catalysed by ornithine decarboxylase (ODC), or indirectly by the decarboxylation of arginine by arginine decarboxylase (ADC), via agmatine. Higher PAs (SPD and SPM) are produced by the sequential addition of aminopropyl moieties to the PUT skeleton through enzymatic reactions catalyzed by the SPD and SPM synthases. S-adenosylmethionine decarboxylase (SAMDC) is responsible for the synthesis of decarboxylated S-adenosylmethionine which is used for the addition of the aminopropyl moiety. Besides the diamine oxidases (DAOs) and PA oxidase (PAOs), which catalyse the terminal catabolism of PAs in the apoplast, PAs also undergo rapid interconvertion in the PA-cycle in the peroxisomes, where the partial and/or full back-conversion of SPM to SPD and of SPD to PUT are catalyse by special PAOs The metabolism of PAs in fungi shows some differences compared to plants. Compare to plants, not all fungi contain SPM. In fungi SPD is the most important PA, which involved in all the functions such as the regulation of cell proliferation and differentiation, vegetative growth, spore germination, sporulation or dimorphic transition, probably in all fungi. Another pronounced difference is, that in plants PUT is synthesized by two alternative pathways catalysed by ODC and ADC, but in fungi, there is a unique pathway PUT formation in which the enzyme ODC plays the central role. The synthesis of higher PAs (SPD and SPM) functions similarly as in plants, as also the terminal catabolism of PAs. Although completed with some additional enzymatic reactions the back-conversion of SPM into SPD, and SPD to PUT is also occurred in fungi.

Due to their cationic nature, they are able to interact with negatively charged macromolecules (such as nucleic acids, phospholipids, proteins) and stabilize their structure, especially under stress conditions. According to these bindings, PAs influence the membrane permeability characteristics, induce changes in protein function. Besides their direct protective role, their binding to the chromatin cause alteration in the availability of genomic sites to DNA or RNA polymerases; furthermore several protein kinases have been shown to be influenced at the transcriptional or post-translational level by PAs. In addition PA metabolism is also related to several other metabolic pathways, such as proline, and the synthesis of plant hormones and signalling molecules, such as ethylene, H_2O_2 or NO. Although several studies have been published on both the changes of PA levels during abiotic stress and protective effects of PAs against various abiotic stress factors, such as drought or cadmium, the mechanism by which PAs act during plant-microbe interaction is sometimes contradictory. In addition, the several types of the interactions between the plants and the microbes resulted in a divers and complex action mechanisms.

PAs are not only essential to maintain cell viability, but function as signalling molecules regulating many of the responses that help the plants to cope with biotic stresses and also to build symbiotic interactions.

Photorespiration

Photorespiration (also known as the oxidative photosynthetic carbon cycle, or C_2

photosynthesis) refers to a process in plant metabolism where the enzyme RuBisCO oxygenates RuBP, wasting some of the energy produced by photosynthesis. The desired reaction is the addition of carbon dioxide to RuBP (carboxylation), a key step in the Calvin–Benson cycle, however approximately 25% of reactions by RuBisCO instead add oxygen to RuBP (oxygenation), creating a product that cannot be used within the Calvin–Benson cycle. This process reduces the efficiency of photosynthesis, potentially reducing photosynthetic output by 25% in C3 plants. Photorespiration involves a complex network of enzyme reactions that exchange metabolites between chloroplasts, leaf peroxisomes and mitochondria.

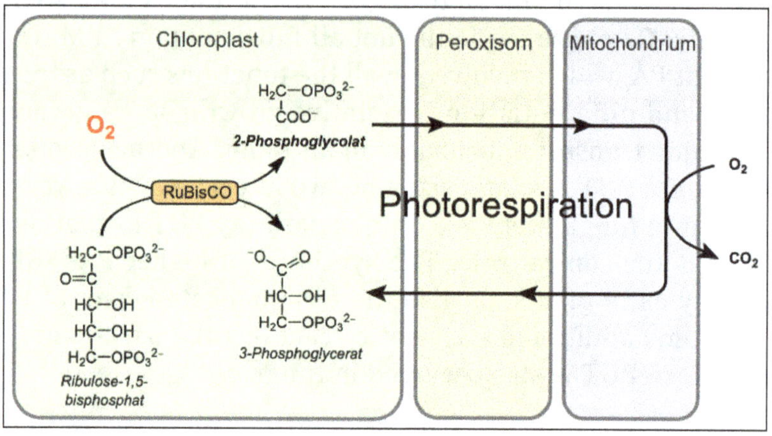

Simplified C_2 cycle.

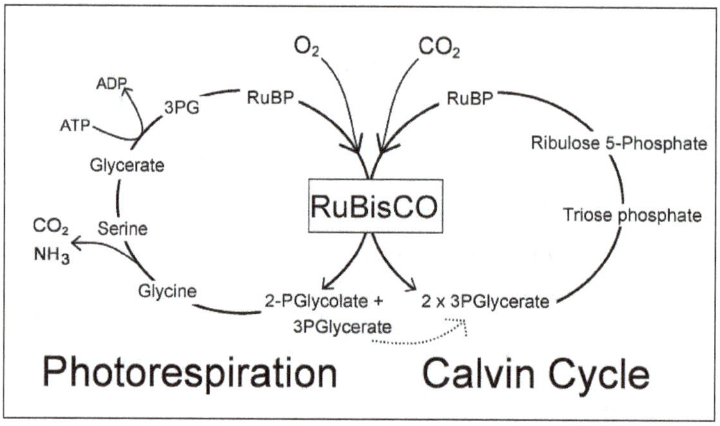

Simplified photorespiration and Calvin cycle.

The oxygenation reaction of RuBisCO is a wasteful process because 3-phosphoglycerate (G3P) is created at a reduced rate and higher metabolic cost compared with RuBP carboxylase activity. While photorespiratory carbon cycling results in the formation of G3P eventually, around 25% of carbon fixed by photorespiration is re-released as CO 2 and nitrogen, as ammonia. Ammonia must then be detoxified at a substantial cost to the cell. Photorespiration also incurs a direct cost of one ATP and one NAD(P)H.

While it is common to refer to the entire process as photorespiration, technically the term refers only to the metabolic network which acts to rescue the products of the oxygenation reaction (phosphoglycolate).

Photorespiratory Reactions

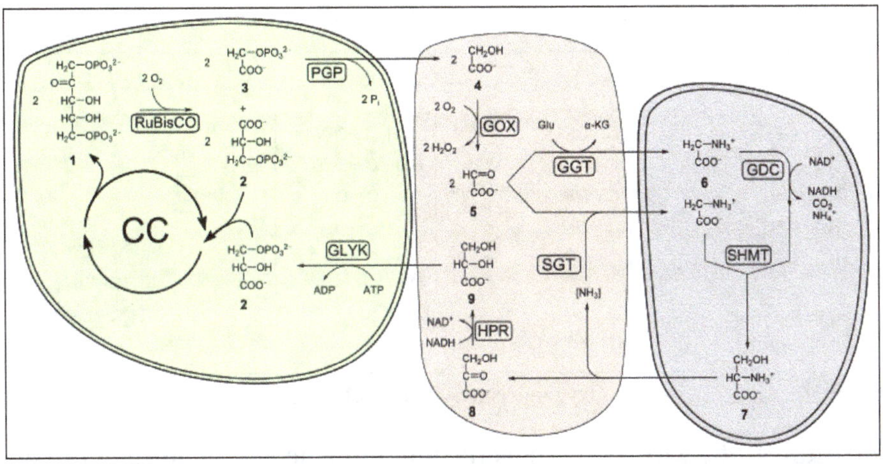

Photorespiration.

Addition of molecular oxygen to ribulose-1, 5-bisphosphate produces 3-phosphoglycerate (PGA) and 2-phosphoglycolate (2PG, or PG). PGA is the normal product of carboxylation, and productively enters the Calvin cycle. Phosphoglycolate, however, inhibits certain enzymes involved in photosynthetic carbon fixation (hence is often said to be an 'inhibitor of photosynthesis'). It is also relatively difficult to recycle: In higher plants it is salvaged by a series of reactions in the peroxisome, mitochondria, and again in the peroxisome where it is converted into glycerate. Glycerate reenters the chloroplast and by the same transporter that exports glycolate. A cost of 1 ATP is associated with conversion to 3-phosphoglycerate (PGA) (Phosphorylation), within the chloroplast, which is then free to re-enter the Calvin cycle.

Several costs are associated with this metabolic pathway; the production of hydrogen peroxide in the peroxisome (associated with the conversion of glycolate to glyoxylate). Hydrogen peroxide is a dangerously strong oxidant which must be immediately split into water and oxygen by the enzyme catalase. The conversion of 2× 2Carbon glycine to 1 C3 serine in the mitochondria by the enzyme glycine-decarboxylase is a key step, which releases CO_2, NH_3, and reduces NAD to NADH. Thus, 1 CO_2 molecule is produced for every 2 molecules of O_2 (two deriving from RuBisCO and one from peroxisomal oxidations). The assimilation of NH_3 occurs via the GS-GOGAT cycle, at a cost of one ATP and one NADPH.

Cyanobacteria have three possible pathways through which they can metabolise 2-phosphoglycolate. They are unable to grow if all three pathways are knocked out, despite having a carbon concentrating mechanism that should dramatically reduce the rate of photorespiration.

Substrate Specificity of RuBisCO

The oxidative photosynthetic carbon cycle reaction is catalyzed by RuBP oxygenase activity:

RuBP + O_2 → Phosphoglycolate + 3-phosphoglycerate + $2H^+$

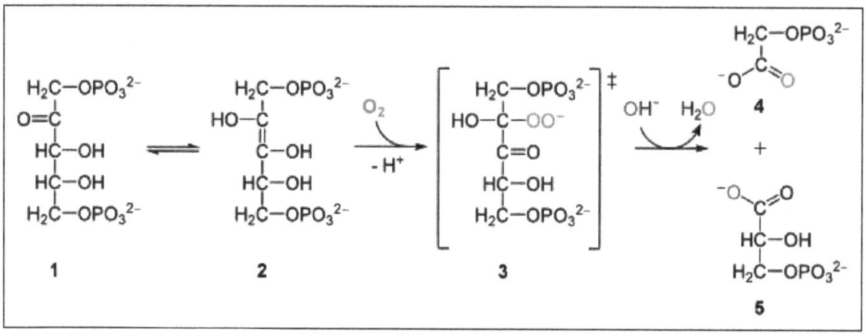

Oxygenase activity of RuBisCO.

During the catalysis by RuBisCO, an 'activated' intermediate is formed (an enediol intermediate) in the RuBisCO active site. This intermediate is able to react with either CO_2 or O_2. It has been demonstrated that the specific shape of the RuBisCO active site acts to encourage reactions with CO_2. Although there is a significant "failure" rate (~25% of reactions are oxygenation rather than carboxylation), this represents significant favouring of CO_2, when the relative abundance of the two gases is taken into account: In the current atmosphere, O_2 is approximately 500 times more abundant, and in solution O_2 is 25 times more abundant than CO_2.

The ability of RuBisCO to specify between the two gases is known as its selectivity factor (or Srel), and it varies between species, with angiosperms more efficient than other plants, but with little variation among the vascular plants.

A suggested explanation into RuBisCO's inability to discriminate completely between CO_2 and O_2 is that it is an evolutionary relic: The early atmosphere in which primitive plants originated contained very little oxygen, the early evolution of RuBisCO was not influenced by its ability to discriminate between O_2 and CO_2.

Conditions which affect Photorespiration

Photorespiration rates are increased by:

- Altered substrate availability (lowered CO_2 or increased O_2): Factors which influence this include the atmospheric abundance of the two gases, the supply of the gases to the site of fixation (i.e. in land plants: Whether the stomata are open or closed), the length of the liquid phase (how far these gases have to diffuse through water in order to reach the reaction site). For example, when the stomata are closed to prevent water loss during drought: This limits the CO_2

supply, while O_2 production within the leaf will continue. In algae (and plants which photosynthesise underwater); gases have to diffuse significant distances through water, which results in a decrease in the availability of CO_2 relative to O_2. It has been predicted that the increase in ambient CO_2 concentrations predicted over the next 100 years may reduce the rate of photorespiration in most plants by around 50%. However, at temperatures higher than the photosynthetic thermal optimum, the increases in turnover rate are not translated into increased CO_2 assimilation because of the decreased affinity of Rubisco for CO_2.

- Increased temperature: At higher temperatures RuBisCO is less able to discriminate between CO_2 and O_2. This is because the enediol intermediate is less stable. Increasing temperatures also reduce the solubility of CO_2, thus reducing the concentration of CO_2 relative to O_2 in the chloroplast.

Biological Adaptation to Minimize Photorespiration

Certain species of plants or algae have mechanisms to reduce uptake of molecular oxygen by RuBisCO. These are commonly referred to as Carbon Concentrating Mechanisms (CCMs), as they increase the concentration of CO_2 so that RuBisCO is less likely to produce glycolate through reaction with O_2.

Biochemical Carbon Concentrating Mechanisms

Biochemical CCMs concentrate carbon dioxide in one temporal or spatial region, through metabolite exchange. C4 and CAM photosynthesis both use the enzyme Phosphoenolpyruvate carboxylase (PEPC) to add CO_2 to a 3-Carbon sugar. PEPC is faster than RuBisCO, and more selective for CO_2.

C4

Maize uses the C4 pathway, minimizing photorespiration.

C4 plants capture carbon dioxide in their mesophyll cells (using an enzyme called phosphoenolpyruvate carboxylase which catalyzes the combination of carbon dioxide with a compound called phosphoenolpyruvate (PEP)), forming oxaloacetate. This oxaloacetate is then converted to malate and is transported into the bundle sheath cells (site of carbon dioxide fixation by RuBisCO) where oxygen concentration is low to avoid photorespiration. Here, carbon dioxide is removed from the malate and combined with RuBP by RuBisCO in the usual way, and the Calvin Cycle proceeds as normal. The CO_2 concentrations in the Bundle Sheath are approximately 10–20 fold higher than the concentration in the mesophyll cells.

This ability to avoid photorespiration makes these plants more hardy than other plants in dry and hot environments, wherein stomata are closed and internal carbon dioxide levels are low. Under these conditions, photorespiration does occur in C4 plants, but at a much reduced level compared with C3 plants in the same conditions. C4 plants include sugar cane, corn (maize), and sorghum.

Cam (Crassulacean Acid Metabolism)

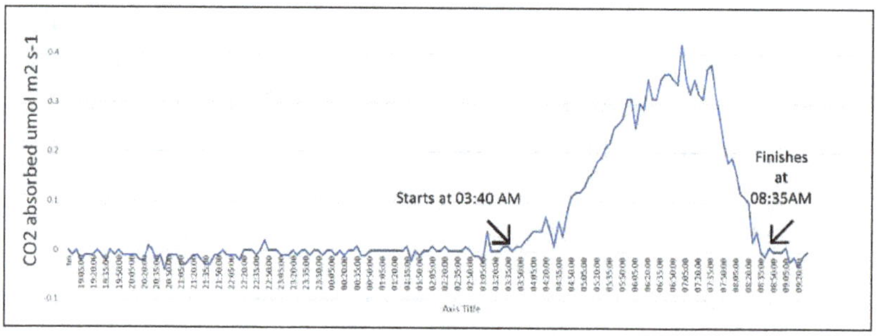

Overnight graph of CO_2 absorbed by a CAM plant.

CAM plants, such as cacti and succulent plants, also use the enzyme PEP carboxylase to capture carbon dioxide, but only at night. Crassulacean acid metabolism allows plants to conduct most of their gas exchange in the cooler night-time air, sequestering carbon in 4-carbon sugars which can be released to the photosynthesizing cells during the day. This allows CAM plants to reduce water loss (transpiration) by maintaining closed stomata during the day. CAM plants usually display other water-saving characteristics, such as thick cuticles, stomata with small apertures, and typically lose around 1/3 of the amount of water per CO_2 fixed.

Algae

There have been some reports of algae operating a biochemical CCM: shuttling metabolites within single cells to concentrate CO2 in one area. This process is not fully understood.

Biophysical Carbon-concentrating Mechanisms

This type of carbon-concentrating mechanism (CCM) relies on a contained compartment within the cell into which CO2 is shuttled, and where RuBisCO is highly expressed.

In many species, biophysical CCMs are only induced under low carbon dioxide concentrations. Biophysical CCMs are more evolutionarily ancient than biochemical CCMs. There is some debate as to when biophysical CCMs first evolved, but it is likely to have been during a period of low carbon dioxide, after the Great Oxygenation Event (2.4 billion years ago). Low CO_2 periods occurred around 750, 650, and 320–270 million years ago.

Eukaryotic Algae

In nearly all species of eukaryotic algae (Chloromonas being one notable exception), upon induction of the CCM, ~95% of RuBisCO is densely packed into a single subcellular compartment: The pyrenoid. Carbon dioxide is concentrated in this compartment using a combination of CO2 pumps, bicarbonate pumps, and carbonic anhydrases. The pyrenoid is not a membrane bound compartment, but is found within the chloroplast, often surrounded by a starch sheath (which is not thought to serve a function in the CCM).

Hornworts

Certain species of hornwort are the only land plants which are known to have a biophysical CCM involving concentration of carbon dioxide within pyrenoids in their chloroplasts.

Cyanobacteria

Cyanobacterial CCMs are similar in principle to those found in eukaryotic algae and hornworts, but the compartment into which carbon dioxide is concentrated has several structural differences. Instead of the pyrenoid, cyanobacteria contain carboxysomes, which have a protein shell, and linker proteins packing RuBisCO inside with a very regular structure. Cyanobacterial CCMs are much better understood than those found in eukaryotes, partly due to the ease of genetic manipulation of prokaryotes.

Possible Purpose of Photorespiration

Reducing photorespiration may not result in increased growth rates for plants. Photorespiration may be necessary for the assimilation of nitrate from soil. Thus, a reduction in photorespiration by genetic engineering or because of increasing atmospheric carbon dioxide (due to fossil fuel burning) may not benefit plants as has been proposed. Several physiological processes may be responsible for linking photorespiration and nitrogen assimilation. Photorespiration increases availability of NADH, which is required for the conversion of nitrate to nitrite. Certain nitrite transporters also transport bicarbonate, and elevated CO_2 has been shown to suppress nitrite transport into chloroplasts. However, in an agricultural setting, replacing the native photorespiration pathway with an engineered synthetic pathway to metabolize glycolate in the chloroplast resulted in a 40 percent increase in crop growth.

Although photorespiration is greatly reduced in C4 species, it is still an essential pathway – mutants without functioning 2-phosphoglycolate metabolism cannot grow in normal conditions. One mutant was shown to rapidly accumulate glycolate.

Although the functions of photorespiration remain controversial, it is widely accepted that this pathway influences a wide range of processes from bioenergetics, photosystem II function, and carbon metabolism to nitrogen assimilation and respiration. The oxygenase reaction of RuBisCO may prevent CO_2 depletion near its active sites and contributes to the regulation of CO_2. Concentration in the atmosphere. The photorespiratory pathway is a major source of hydrogen peroxide (H_2O_2) in photosynthetic cells. Through H_2O_2 production and pyrimidine nucleotide interactions, photorespiration makes a key contribution to cellular redox homeostasis. In so doing, it influences multiple signalling pathways, in particular, those that govern plant hormonal responses controlling growth, environmental and defense responses, and programmed cell death.

It has been postulated that photorespiration may function as a "safety valve", preventing the excess of reductive potential coming from an overreduced NADPH-pool from reacting with oxygen and producing free radicals, as these can damage the metabolic functions of the cell by subsequent oxidation of membrane lipids, proteins or nucleotides. The mutants deficient in photorespiratory enzymes are characterized by a high redox level in the cell, impaired stomatal regulation, and accumulation of formate.

SECONDARY METABOLIC PATHWAYS

Secondary metabolism produces a large number of specialized compounds (estimated 200,000) that do not aid in the growth and development of plants but are required for the plant to survive in its environment. Secondary metabolism is connected to primary metabolism by using building blocks and biosynthetic enzymes derived from primary metabolism. Primary metabolism governs all basic physiological processes that allow a plant to grow and set seeds, by translating the genetic code into proteins, carbohydrates, and amino acids. Specialized compounds from secondary metabolism are essential for communicating with other organisms in mutualistic (e.g. attraction of beneficial organisms such as pollinators) or antagonistic interactions (e.g. deterrent against herbivores and pathogens). They further assist in coping with abiotic stress such as increased UV-radiation. The broad functional spectrum of specialized metabolism is still not fully understood. In any case, a good balance between products of primary and secondary metabolism is best for a plant's optimal growth and development as well as for its effective coping with often changing environmental conditions. Well known specialized compounds include alkaloids, polyphenols including flavonoids, and terpenoids. Humans use quite a lot of these compounds, or the plants from which they originate, for culinary, medicinal and nutraceutical purposes.

Research into secondary plant metabolism primarily took off in the later half of the 19th century, however, there was still much confusion over what the exact function and usefulness of these compounds were. All that was known was that secondary plant metabolites were "by-products" of the primary metabolism and were not crucial to the plant's survival. Early research only succeeded as far as categorizing the secondary plant metabolites but did not give real insight into the actual function of the secondary plant metabolites. The study of plant metabolites is thought to have started in the early 1800s when Friedrich Willhelm Serturner isolated morphine from opium poppy, and after that new discoveries were made rapidly. In the early half of the 1900s, the main research around secondary plant metabolism was dedicated to the formation of secondary metabolites in plants, and this research was compounded by the use of tracer techniques which made deducing metabolic pathways much easier. However, there was still not much research being conducted into the functions of secondary plant metabolites until around the 1980s. Before then, secondary plant metabolites were thought of as simply waste products. In the 1970s, however, new research showed that secondary plant metabolites play an indispensable role in the survival of the plant in its environment. One of the most ground breaking ideas of this time argued that plant secondary metabolites evolved in relation to environmental conditions, and this indicated the high gene plasticity of secondary metabolites, but this theory was ignored for about half a century before gaining acceptance. Recently, the research around secondary plant metabolites is focused around the gene level and the genetic diversity of plant metabolites. Biologists are now trying to trace back genes to their origin and re-construct evolutionary pathways.

Primary vs. Secondary Plant Metabolism

Primary metabolism in a plant comprises all metabolic pathways that are essential to the plant's survival. Primary metabolites are compounds that are directly involved in the growth and development of a plant whereas secondary metabolites are compounds produced in other metabolic pathways that, although important, are not essential to the functioning of the plant. However, secondary plant metabolites are useful in the long term, often for defense purposes, and give plants characteristics such as color. Secondary plant metabolites are also used in signalling and regulation of primary metabolic pathways. Plant hormones, which are secondary metabolites, are often used to regulate the metabolic activity within cells and oversee the overall development of the plant. As mentioned above in the History tab, secondary plant metabolites help the plant maintain an intricate balance with the environment, often adapting to match the environmental needs. Plant metabolites that color the plant are a good example of this, as the coloring of a plant can attract pollinators and also defend against attack by animals.

Types of Secondary Metabolites in plants

There is no fixed, commonly agreed upon system for classifying secondary metabolites. Based on their biosynthetic origins, plant secondary metabolites can be divided into three major groups:

- Flavonoids and allied phenolic and polyphenolic compounds.
- Terpenoids.
- Nitrogen-containing alkaloids and sulphur-containing compounds.

Other researchers have classified secondary metabolites into following, more specific types:

Class	Type	Number of known metabolites	Examples
Alkaloids	Nitrogen-containing	21000	Cocaine, Psilocin, Caffeine, Nicotine, Morphine, Berberine, Vincristine, Reserpine, Galantamine, Atropine, Vincamine, Quinidine, Ephedrine, Quinine
Non-protein amino acids (NPAAs)	Nitrogen-containing	700	NPAAs are produced by specific plant families such as Leguminosae, Cucurbitaceae, Sapindaceae, Aceraceae and Hippocastanaceae. Examples: Azatyrosine, Canavanine
Amines	Nitrogen-containing	100	
Cyanogenic glycosides	Nitrogen-containing	60	Amygdalin, Dhurrin, Linamarin, Lotaustralin, Prunasin
Glucosinolates	Nitrogen-containing	100	
Alkamides	Nitrogen-containing	150	
Lectins, peptides and polypeptides	Nitrogen-containing	2000	Concanavalin A
Terpenes	Without nitrogen	>15,000	Azadirachtin, Artemisinin, Tetrahydrocannabinol
Steroids and saponins	Without nitrogen	NA	These are terpenoids with a particular ring structure. Cycloartenol
Flavonoids and Tannins	Without nitrogen	5000	Luteolin, tannic acid
Phenylpropanoids, lignins, coumarins and lignans	Without nitrogen	2000	Resveratrol
Polyacetylenes, fatty acids and waxes	Without nitrogen	1500	
Polyketides	Without nitrogen	750	
Carbohydrates and organic acids	Without nitrogen	200	

Cyanogenic Glycoside

Many plants have adapted to iodine-deficient terrestrial environment by removing iodine from their metabolism, in fact iodine is essential only for animal cells. An important antiparasitic action is caused by the block of the transport of iodide of animal cells inhibiting sodium-iodide symporter (NIS). Many plant pesticides are cyanogenic glycoside which liberate cyanide, which, blocking cytochrome c oxidase and NIS, is poisonous only for a large part of parasites and herbivores and not for the plant cells in which it seems useful in seed dormancy phase. To get a better understanding of how secondary metabolites play a big role in plant defense mechanisms we can focus on the recognizable defense-related secondary metabolites, cyanogenic glycosides. The compounds of these secondary metabolites are found in over 2000 plant species. Its structure allows the release of cyanide, a poison produced by certain bacteria, fungi, and algae that is found in numerous plants. Animals and humans possess the ability to detoxify cyanide from their systems naturally. Therefore, cyanogenic glycosides can be used for positive benefits in animal systems always. For example, the larvae of the southern armyworm consumes plants that contain this certain metabolite and have shown a better growth rate with this metabolite in their diet, as opposed to other secondary metabolite-containing plants. Although this example shows cyanogenic glycosides being beneficial to the larvae many still argue that this metabolite can do harm. To help in determining whether cyanogenic glycosides are harmful or helpful researchers look closer at its biosynthetic pathway. Past research suggests that cyanogenic glucosides stored in the seed of the plant are metabolized during germination to release nitrogen for seedling to grow. With this, it can be inferred that cyanogenic glycosides play various roles in plant metabolism. Though subject to change with future research, there is no evidence showing that cyanogenic glycosides are responsible for infections in plants.

Phytic Acid

Phytic acid is the main method of phosphorus storage in plant seeds, but is not readily absorbed by many animals (only absorbed by ruminant animals). Not only is phytic acid a phosphorus storage unit, but it also is a source of energy and cations, a natural antioxidant for plants, and can be a source of myoinositol which is one of the preliminary pieces for cell walls.

Phytic acid is also known to bond with many different minerals, and by doing so prevents those minerals from being absorbed; making phytic acid an anti-nutrient. There is a lot of concern with phytic acids in nuts and seeds because of its anti-nutrient characteristics. In preparing foods with high phytic acid concentrations, it is recommended they be soaked in after being ground to increase the surface area. Soaking allows the seed to undergo germination which increases the availability of vitamins and nutrient, while reducing phytic acid and protease inhibitors, ultimately increasing the nutritional value. Cooking can also reduce the amount of phytic acid in food but soaking is much more effective.

Phytic acid is an antioxidant found in plant cells that most likely serves the purpose of preservation. This preservation is removed when soaked, reducing the phytic acid and allowing the germination and growth of the seed. When added to foods it can help prevent discoloration by inhibiting lipid peroxidation. There is also some belief that the chelating of phytic acid may have potential use in the treatment of cancer.

Carotenoids

Carotenoids are organic pigments found in the chloroplasts and chromoplasts of plants. They are also found in some organisms such as algae, fungi, some bacteria, and certain species of aphids. There are over 600 known carotenoids. They are split into two classes, xanthophylls and carotenes. Xanthophylls are carotenoids with molecules containing oxygen, such as lutein and zeaxanthin. Carotenes are carotenoids with molecules that are unoxygenated, such as α-carotene, β-carotene and lycopene. In plants, carotenoids can occur in roots, stems, leaves, flowers, and fruits. Carotenoids have two important functions in plants. First, they can contribute to photosynthesis. They do this by transferring some of the light energy they absorb to chlorophylls, which then uses this energy for photosynthesis. Second, they can protect plants which are over-exposed to sunlight. They do this by harmlessly dissipating excess light energy which they absorb as heat. In the absence of carotenoids, this excess light energy could destroy proteins, membranes, and other molecules. Some plant physiologists believe that carotenoids may have an additional function as regulators of certain developmental responses in plants. Tetraterpenes are synthesized from DOXP precursors in plants and some bacteria. Carotenoids involved in photosynthesis are formed in chloroplasts; Others are formed in plastids. Carotenoids formed in fungi are presumably formed from mevalonic acid precursors. Carotenoids are formed by a head-to-head condensation of geranylgeranyl pyrophosphate or diphosphate (GGPP) and there is no NADPH requirement.

Atropine

Atropine is an alkaloid derived from the solanaceous plants Atropa belladonna (deadly nightshade), Hyoscyamus niger (black henbane), and Datura stramonium (thornapple). These plants contain a mixture of two closely related alkaloids, hyoscyamine and hyoscine; atropine is a mixture of two isomers of hyoscyamine. In 1867, von Bezold found that atropine blocked the slowing of the heart caused by vagal stimulation. We now know that atropine blocks the action of the neurotransmitter acetylcholine at all the nerve endings where the membrane receptors are of the so-called muscarinic type. This includes those of the parasympathetic nervous system in the heart, glandular tissue, and smooth muscle. Thus atropine causes a rise in heart rate and inhibits secretions (for example of saliva, causing a dry mouth, and of the digestive enzymes). It also relaxes smooth muscle in the gastrointestinal tract, the urinary bladder, and the bronchial trees, by preventing the effects of the normal background discharge of parasympathetic neurons to these organs.

The central nervous system also contains muscarinic receptors. Blockade of these by atropine leads to restlessness and mental excitement, and can improve the rigidity and tremor characteristic of Parkinson's disease. Large doses of atropine can cause hallucination.

Long-lasting pupillary dilation results if atropine drops are placed in the eye. The iris has both circular and radial muscles, and the balance between the tonic activities of these two muscle groups controls the pupil diameter. The circular muscle is under parasympathetic control, so when the transmitter, acetylcholine, is blocked with atropine, the pupil will dilate.

Flavonoids

Flavonoids are compounds that are secondary metabolites, but which play an important role in the biological activities of plants. They can be responsible for the color of flowers and fruits and for the attraction of pollinators. They also participate in plant-microorganism symbiosis. These relationships can be used to naturally control weeds and insect pests and reduce stress and diseases in order to increase crop yield. To improve the understanding of the different biological systems where flavonoids are involved in their symbiotic relationships and in plant physiology, tools such as metabolomics are used, which give a broader picture and allow to search for strategies to solve problems specific to the agricultural sector.

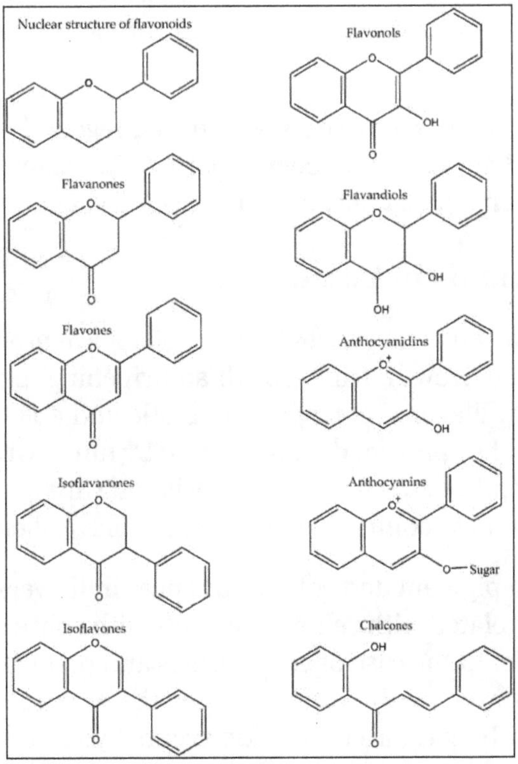

Structure of the different types of flavonoids.

Flavonoids are secondary metabolites of plants; this group of phenolic compounds includes approximately 4500 compounds. They are classified into different subgroups. Their nuclear structure includes carbons C6-C3-C6, and the diversity of flavonoids depends on the position of the aromatic ring.

Flavonoids and flavones are the most common in plants, whereas flavanones, flavanols, dihydroflavones, and dihydrochalcones have a limited distribution. Flavonoids are present in plant tissues in relatively high concentrations in free forms (aglycones) or conjugated with sugar molecules (glycosides).

Flavonoids are polyphenolic compounds found in all vascular and non-vascular plants. They are important in the diet of humans because they possess a high nutritional value; besides, they are used as effective medicines in the treatments of certain diseases, and therefore, they are called nutraceuticals. Nowadays, a diet rich in fruits and vegetables is recommended to contribute to the prevention and treatment of cardiovascular diseases, diabetes, cancer, chronic inflammatory disorders, and degenerative diseases.

Some examples of nutraceutical flavonoids are the following: isoflavones (genistein, daidzein) obtained from celery, soybeans, and other legumes. Isoflavones have antitumor, anticancer, and antioxidant action, and they improve immune response, lower cardiovascular risk, and menopausal symptoms. Quercetin is contained in onion, citrus fruits, broccoli, red grapes, apple, and cherries. Kaempferol is present in broccoli and radishes. Anthocyanins are present in red wine and fruits. These flavonoids are important due to their antioxidant, anticancer, and antithrombotic properties, as well as their ability to lower blood cholesterol.

Their concentration can vary among species: In fruits, vegetables, and medicinal plants, ranges from 0 to 6125.6 mg kg−1 have been reported. For example, 1720.5 mg kg−1 has been observed in spinach, 3575.4 mg kg−1 in strawberry, and 459.9 mg kg−1 in apple.

Function of Flavonoids in Plants

In plants, flavonoids play important roles in many biological processes. They participate in seed development and growth, fruit growth and ripening, pollen tube germination, and hormone transport. Flavonoids respond to biotic and abiotic factors by providing antioxidant properties; they prevent damage caused by fungi, viruses, bacteria, and herbivores; function as chemical messengers in association with mycorrhizae and bacteria; act as chemical attractants to pollinating animals; and have allelopathic functions.

They also participate in pigment and color differences in flowers, fruits, and seeds. For example, flavonols are related with yellow, flavanols with ochre to brown, and anthocyanins with red to purple. In the case of corn kernels and petunia flowers, anthocyanins and proanthocyanidins are mainly responsible for the pigments. These accumulate in the vacuole or cell wall. In corn, accumulation occurs by vacuolar sequestration of anthocyanins.

As part of their defense strategy, plants induce systems of antioxidant activity, reactive oxygen species (ROS), enzymatic and non-enzymatic, soluble in water and in lipids, located in different cell compartments. The enzymatic ROS system consists of several enzymes such as superoxide dismutase (SOD), catalase (CAT), ascorbate peroxidase (APX), monodehydroascorbate reductase (MDAR), dehydroascorbate reductase (DHAR), glutathione peroxidase (GPX), and glutathione reductase (GR). Non-enzymatic antioxidants include pigments (carotenoids, anthocyanidins), vitamins (A, C, E), and flavonoids, among others. The synthesis of the latter compounds by the plants is related to environmental biotic and abiotic stress factors. Plants subjected to conditions of severe stress accumulate dihydroxy B-ring substituted flavonoids, which are effective eliminators of ROS.

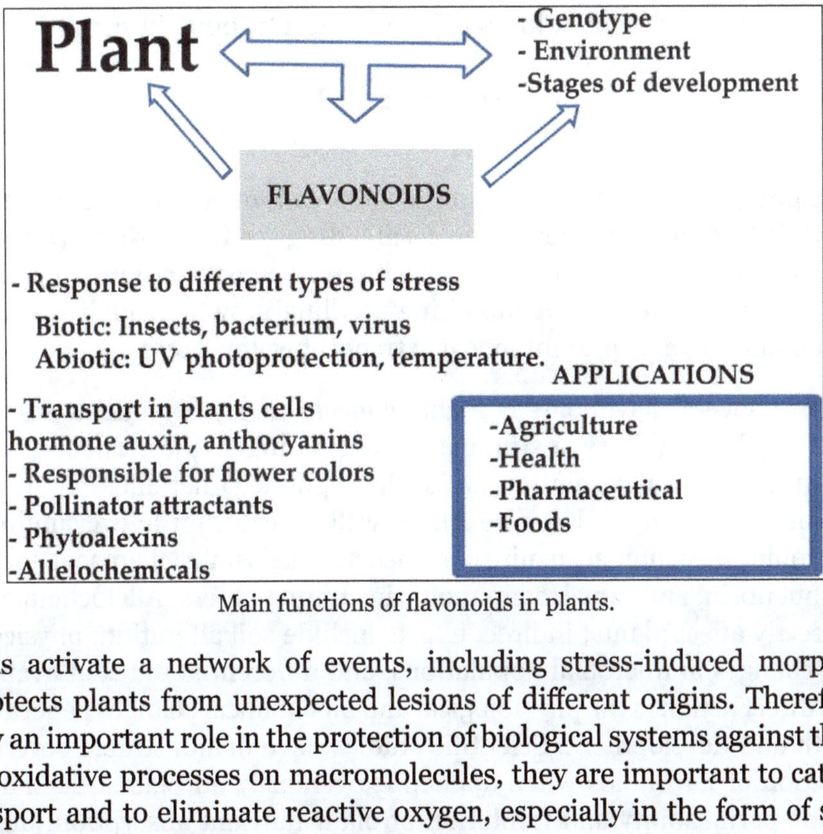

Main functions of flavonoids in plants.

Flavonoids activate a network of events, including stress-induced morphogenesis, which protects plants from unexpected lesions of different origins. Therefore, flavonoids play an important role in the protection of biological systems against the harmful effects of oxidative processes on macromolecules, they are important to catalyze electron transport and to eliminate reactive oxygen, especially in the form of superoxide anions, hydroxyl radicals, lipid peroxides, or hydroperoxides. In this way, they block the deleterious activity of these substances on the cells.

Flavonoids: Allelopathy and its Applications in Agriculture

Allelopathy is defined as the direct or indirect effect of secondary compounds produced by a donor plant on a recipient plant. This type of relationship can be beneficial or harmful.

Currently, to solve multiple problems in agriculture, allelopathy is being considered as a natural control of weeds and insect pests and to reduce stress and diseases, in order to increase crop yield. Weeds are the species that compete the most with crops, and for their management, aqueous extracts are used, such as natural herbicides from sorghum, sunflower, eucalyptus, and rice, among others; a greater efficacy is obtained when they are mixed together than alone. On the other hand, it has been demonstrated that exudates of roots of rice plants reduce the attack of fungi of the genus Fusarium in melon. In addition, plants of Brassica napus L., incorporated into the soil, decrease the population of certain nematodes in orchards. These examples show the potential of some allelochemicals to generate herbicides, fungicides, insecticides, and nematicides.

Knowledge of allelopathy can be valuable to improve crop rotation proposals. Recent publications highlight the role of flavonoids in allelopathy, involved in soil interactions, since they have been identified in significant concentrations in many bioactive root exudates. Simple phenols and flavonoids are released by the decomposition of plant tissues as leachates and by the process of microbial degradation and transformation in the soil.

In many legumes, it has been shown that the flavonoids quercetin and kaempferol, free and glycosylated, commonly released by germinating seeds and roots, persist for days in the soil and possess an important phytoinhibitory activity, stimulating seed germination at low concentrations but inhibiting seedling growth at high concentrations. These compounds are also present in leaf extracts of walnut trees.

The activity of several flavonoids is highly concentration-dependent; some of these compounds may be inhibitory or stimulatory, depending on the availability of the concentration in the soil/water solution in the rhizosphere. Other allelopathic flavonoids are lutonarin, saponarin, and isovitexin, as well as catechin and cyanidin, luteolin 7-Ob-glucuronide, neochamaejasmin, mesoneochamaejasmin, chamaejasmenin, genkwanol, daphnodorin, and dihydrodaphnodorin, among others. Allelochemicals directly and indirectly affect plants; indirect effects include soil alteration, physicochemical properties, changes in microbial populations, and differential nutrient availability for plants. The direct action is the physiological and biochemical changes generated during plant growth and development. Scientific studies have demonstrated several mechanisms of action of flavonoids when applied exogenously on plants. Such as changes in membrane permeability and inhibition of plant nutrient absorption, inhibition of cell division, elongation and submicroscopic structure, effects on photosynthesis and respiration of the plant, consequences on different enzymatic functions and activities, effects on the synthesis of endogenous hormones and proteins, and the disruption of adenosine triphosphate (ATP) formation.

Metabolomics: Flavonoids as a Tool in Agriculture

It has been described that over 7000 natural products belong to the flavonoid family and to analyze them, physicochemical methods as nuclear magnetic resonance (NMR) and mass spectrometry have been used, but nowadays the most important issue is concerned with the identification of free and conjugated forms of these compounds. Metabolite profiling is an essential tool to analyze the effects of pathway engineering approaches; in this sense, the metabolomics approach used to solve this problem as the liquid-chromatography-mass spectrometry has supported the quantitative and qualitative analysis of flavonoids. In addition, straightforward and efficient methods approach toward rapid flavonoid identification by combining simple high-performance liquid chromatography (HPLC) and NMR methods, facilitating the analysis of flavones and flavanones.

Flavonoids are a diverse group in agricultural crops, and there are studies using metabolomic tools to analyze the content of flavonoids using different analytical techniques.

Red tomato contains flavonoids mainly in the peel; through omic studies, it may be possible to modify the pattern of flavonoids in the pulp. Metabolic profiling represents a useful tool to characterize varieties with functional markers, such as flavonoids. For example, metabolite profiling allowed the characterization of Italian tomato landraces and different fruit types. Another study related with metabolomic profiling is the analysis of flavonoid distribution in three Momordica species, where 13 flavonoids were found in a special pattern.

The novel approach of genetic metabolomics referred to as metabolite profiling combined with quantitative trait locus (QTL) analysis was applied to detect flux control points in flavonoids biosynthesis of Populus. It was found that flavonoid profile can be used for QTL analysis to reveal loci that control the flux of their biosynthesis.

Plant metabolism is disrupted by several types of stress. Flavonoids are involved in the response toward abiotic stress; a good model to understand how flavonoids contribute to the mitigation of oxidative and drought stress is Arabidopsis thaliana. Another example is the metabolomic analysis of tea, where the effects of light intensity and temperature on the metabolites in tea grown in the shade were evaluated; they found that most flavonoids (flavan-3-ols, flavonols, and their glucosides) decreased significantly in the shading treatments. Their study also showed a greater effect of temperature on galloylation of catechins than light intensity.

The growing stage of plants has been studied in connection with flavonoid synthesis, an interesting metabolomic study is the analysis of soybean leaves, it showed significant changes in the content of flavonoids and isoflavonoids, and kaempferol derivatives were used as markers. The results demonstrated that metabolite production changed depending on the growing stage, and they mention that the information can be useful to understand physiological characterization and suggest an optimal harvesting time of this crop.

Metabolic analysis has the potential to generate a complete vision of metabolic networks and has revealed multiple detection, quantification, and analysis strategies to evaluate numerous metabolites such as flavonoids. This tool attempts to integrate compound and metabolite analyses along with other biological data overview of plants, for example, phenotypic, morphological, and genetic data.

Symbiosis Flavonoids-microorganisms

Flavonoids are considered signaling compounds in plant-microorganisms symbiosis; their function is signaling in response to pathogens, bacteria, fungi. Their participation is important in the nodulation of roots, where they are secreted to the rhizosphere, in the case of legumes under low nitrogen conditions, and where their interaction is specific with gram-negative bacteria called rhizobia that are nitrogen fixers. They are also involved in the transcription of genes for the biosynthesis of rhizobial signaling molecules called Nod factors, which are perceived by the plant to allow symbiotic root infection. Therefore, they can stimulate or inhibit rhizobial Nod gene expression. Root exudates can be flavonoids that participate as signaling compounds in the arbuscular mycorrhizal symbiosis. Root exudation of flavonoids increases or decreases depending on the response to the symbiotic and pathogenic interaction of the plant and microbes.

It has been shown that certain flavonoids are stimulants for the germination of ectomycorrhizal fungi spores. Depending on the arbuscular mycorrhizal fungi colonizers of the roots and the stage of symbiosis development.

The flavonoid role in the rhizosphere is important in agriculture because flavonoid interaction in the symbiosis with microorganisms can be a tool for nitrogen fixation in soils poor in macronutrients. It can be a strategy to stimulate beneficial bacteria or inhibit harmful bacteria and fungi. It can also be considered in the genetic improvement of plant species.

Table: Role of flavonoids in the rhizosphere.

Participation	Flavonoid	Species	Symbiosis type
Stimulant for germination of spores	Coumestrol, medicarpin, ononin, formononetin, daidzein, genistein, biochanin A and 4′,7-dihydroxyflavanone, 4,4′-dihydroxy-2′-methoxychalcone	Medicago sativa L. cv. Sitel	Fungus
	hesperidin, morin, rutin, quercetin, naringenin, genistein, and chrysin	Suillus bovinus	Fungus
Chemo attractant response from the rhizobia	Luteolin and apigenin	–	Bacterium
Auxin transport inhibitors	Kaempferol	Medicago truncatula	–

Importance of Flavonoids in Pest Control

Every day, the demand for natural products obtained from plants increases to be used as pest control agents. New pesticides are being developed using flavonoids, as they are an alternative to synthetic pesticides. They can inhibit enzymatic activity and prevent the growth of larvae of different insect species. Some flavonoids interfere in the process of moulting and reproduction of several insects, that is, they inhibit the formation of juvenile hormone (ecdysone). Flavonoids inhibit transcription of ecdysone receptor-dependent genes (EcR).

It has been reported that some types of flavonoids have had an effect on agricultural pests with ovicidal effect, oviposition, fecundity, mortality, weight reduction, and emergence of adults. Quercetin, rutin, and naringin showed positive effects for the control of nymphs and adults of the aphid Eriosoma lanigerum Hausmann. These can be used as an insecticide in the integrated management programs of this aphid. Some flavonoids can influence agricultural pests depending on the concentration applied; if they are low, they do not affect them; therefore, it is necessary to test the minimum concentrations for the flavonoid to have an insecticidal effect.

Phytoestrogen

A phytoestrogen is a plant-derived xenoestrogen not generated within the endocrine system, but consumed by eating phytoestrogenic plants. Also called a "dietary estrogen", it is a diverse group of naturally occurring nonsteroidal plant compounds that, because of its structural similarity with estradiol (17-β-estradiol), have the ability to cause estrogenic and/or antiestrogenic effects. Phytoestrogens are not essential nutrients because their absence from the diet does not cause a disease, nor are they known to participate in any normal biological function.

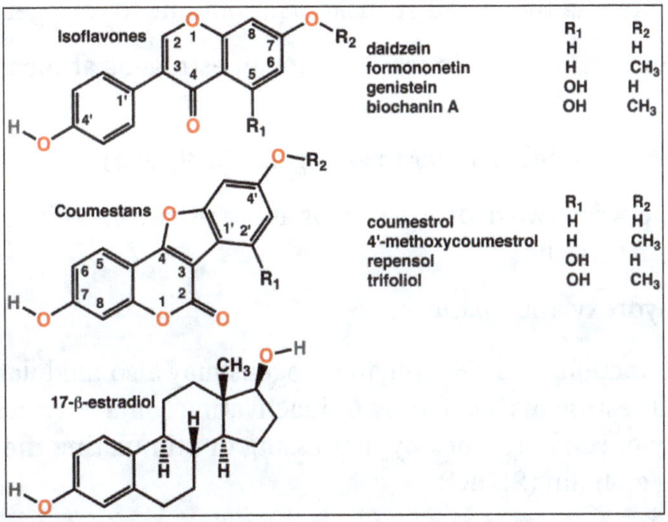

Chemical structures of the most common phytoestrogens found in plants (top and middle) compared with estrogen (bottom) found in animals.

The similarities, at molecular level, of an estrogen and a phytoestrogen allow them to mildly mimic and sometimes act as an antagonist of estrogen. Phytoestrogens were first observed in 1926, but it was unknown if they could have any effect in human or animal metabolism. In the 1940s and early 1950s, it was noticed that some pastures of subterranean clover and red clover (phytoestrogen-rich plants) had adverse effects on the fecundity of grazing sheep.

Structure

Phytoestrogens mainly belong to a large group of substituted natural phenolic compounds: The coumestans, prenylflavonoids and isoflavones are three of the most active in estrogenic effects in this class. The best-researched are isoflavones, which are commonly found in soy and red clover. Lignans have also been identified as phytoestrogens, although they are not flavonoids. Mycoestrogens have similar structures and effects, but are not components of plants; these are mold metabolites of Fusarium, especially common on cereal grains, but also occurring elsewhere, e.g. on various forages. Although mycoestrogens are rarely taken into account in discussions about phytoestrogens, these are the compounds that initially generated the interest on the topic.

Mechanism of Action

Phytoestrogens exert their effects primarily through binding to estrogen receptors (ER). There are two variants of the estrogen receptor, alpha (ER-α) and beta (ER-β) and many phytoestrogens display somewhat higher affinity for ER-β compared to ER-α.

The key structural elements that enable phytoestrogens to bind with high affinity to estrogen receptors and display estradiol-like effects are:

- The phenolic ring that is indispensable for binding to estrogen receptor.
- The ring of isoflavones mimicking a ring of estrogens at the receptors binding site.
- Low molecular weight similar to estrogens (MW=272).
- Distance between two hydroxyl groups at the isoflavones nucleus similar to that occurring in estradiol.
- Optimal hydroxylation pattern.

In addition to interaction with ERs, phytoestrogens may also modulate the concentration of endogenous estrogens by binding or inactivating some enzymes, and may affect the bioavailability of sex hormones by depressing or stimulating the synthesis of sex hormone-binding globulin (SHBG).

Emerging evidence shows that some phytoestrogens bind to and transactivate peroxisome proliferator-activated receptors (PPARs). In vitro studies show an activation of PPARs at concentrations above 1 μM, which is higher than the activation level of ERs. At the concentration below 1 μM, activation of ERs may play a dominant role. At higher concentrations (>1 μM), both ERs and PPARs are activated. Studies have shown that both ERs and PPARs influence each other and therefore induce differential effects in a dose-dependent way. The final biological effects of genistein are determined by the balance among these pleiotrophic actions.

Ecology

These compounds in plants are an important part of their defense system, mainly against fungi.

Phytoestrogens are ancient naturally occurring substances, and as dietary phytochemicals they are considered as co-evolutive with mammals. In the human diet, phytoestrogens are not the only source of exogenous estrogens. Xenoestrogens (novel, man-made), are found as food additives and ingredients, and also in cosmetics, plastics, and insecticides. Environmentally, they have similar effects as phytoestrogens, making it difficult to clearly separate the action of these two kind of agents in studies done on populations.

Avian Studies

The consumption of plants with unusual content of phytoestrogens, under drought conditions, has been shown to decrease fertility in quail. Parrot food as available in nature has shown only weak estrogenic activity. Studies have been conducted on screening methods for environmental estrogens present in manufactured supplementary food, with the purpose of aiding reproduction of endangered species.

Food Sources

According to one study of nine common phytoestrogens in a Western diet, foods with the highest relative phytoestrogen content were nuts and oilseeds, followed by soy products, cereals and breads, legumes, meat products, and other processed foods that may contain soy, vegetables, fruits, alcoholic, and nonalcoholic beverages. Flax seed and other oilseeds contained the highest total phytoestrogen content, followed by soybeans and tofu. The highest concentrations of isoflavones are found in soybeans and soybean products followed by legumes, whereas lignans are the primary source of phytoestrogens found in nuts and oilseeds (e.g. flax) and also found in cereals, legumes, fruits and vegetables. Phytoestrogen content varies in different foods, and may vary significantly within the same group of foods (e.g. soy beverages, tofu) depending on processing mechanisms and type of soybean used. Legumes (in particular soybeans), whole grain cereals, and some seeds are high in phytoestrogens.

A more comprehensive list of foods known to contain phytoestrogens includes:

- Soybeans and soy products
- Tempeh
- Linseed (flax)
- Sesame seeds
- Wheat berries
- Fenugreek (contains diosgenin, but also used to make Testofen, a compound taken by men to increase testosterone)
- Oats Barley

- Beans
- Lentils
- Yams
- Rice
- Alfalfa
- Mung beans
- Apples
- Carrots
- Pomegranates
- Wheat germ
- Rice bran
- Lupin
- Kudzu
- Coffee
- Licorice root
- Mint
- Ginseng
- Hops
- Bourbon whiskey
- Beer
- Fennel
- Anise
- Red clover (sometimes a constituent of green manure)
- Spinach

An epidemiological study of women in the United States found that the dietary intake of phytoestrogens in healthy post-menopausal Caucasian women is less than one milligram daily.

Effects on Humans

In human beings, phytoestrogens are readily absorbed into the circulatory system, circulate in plasma, and are excreted in the urine. Metabolic influence is different from that of grazing animals due to the differences between ruminant versus monogastric digestive systems.

A wide range of beneficial effects of phytoestrogens on the cardiovascular, metabolic, central nervous systems as well as reduction of risk of cancer and post menopausal symptoms have been claimed. However, there is also concern that phytoestrogens may act as endocrine disruptors that adversely affect health. Based on currently available evidence, it is not clear whether the potential health benefits of phytoestrogens outweigh their risks.

Males

It is unclear if phytoestrogens have any effect on male sexuality, with conflicting results about the potential effects of isoflavones originating from soy. A 2010 meta-analysis of fifteen placebo-controlled studies led by Dr. Mark Messina, executive director of the

Soy Nutrition Institute, said that "neither soy foods nor isoflavone supplements alter measures of bioavailable testosterone concentrations in men." Some studies showed that isoflavone supplementation had no effect on sperm concentration, count, or motility, and had no effects on testicular or ejaculate volume. Sperm count decline and increasing rate of testicular cancers in the West may be linked to a higher presence of isoflavone phytoestrogens in the diet, but such a link has not been definitively proven. Furthermore, there is some evidence that phytoestrogens may affect male fertility, but "further investigation is needed before a firm conclusion can be drawn".

Females

It is unclear if phytoestrogens have any effect on the cause or prevention of cancer in females. Some epidemiological studies have suggested a protective effect against breast cancer. Other epidemiological studies found that consumption of soy estrogens is safe for patients with breast cancer, and that it may decrease mortality and recurrence rates. It remains unclear if phytoestrogens can minimize some of the deleterious effects of low estrogen levels (hypoestrogenism) resulting from oophorectomy, menopause, or other causes. A Cochrane Review of the use of phytoestrogens to relieve the vasomotor symptoms of menopause (hot flashes) stated that there was no conclusive evidence to suggest any benefit to their use, although genistein effects should be further investigated.

Infant Formula

Some studies have found that some concentrations of isoflavones may have effects on intestinal cells. At low doses, genistein acted as a weak estrogen and stimulated cell growth; at high doses, it inhibited proliferation and altered cell cycle dynamics. This biphasic response correlates with how genistein is thought to exert its effects. Some reviews express the opinion that more research is needed to answer the question of what effect phytoestrogens may have on infants, but their authors did not find any adverse effects. Studies conclude there are no adverse effects in human growth, development, or reproduction as a result of the consumption of soy-based infant formula compared to conventional cow-milk formula. The American Academy of Pediatrics states: "although isolated soy protein-based formulas may be used to provide nutrition for normal growth and development, there are few indications for their use in place of cow milk-based formula. These indications include (a) for infants with galactosemia and hereditary lactase deficiency (rare) and (b) in situations in which a vegetarian diet is preferred."

Ethnopharmacology

In some countries, phytoestrogenic plants have been used for centuries in the treatment of menstrual and menopausal problems, as well as for fertility problems. Plants used that have been shown to contain phytoestrogens include Pueraria mirifica, and its close relative, kudzu, Angelica, fennel and anise. In a rigorous study, the use of one such source of phytoestrogen, red clover, has been shown to be safe, but ineffective in relieving menopausal symptoms (black cohosh is also used for menopausal symptoms, but does not contain phytoestrogens).

Gossypol

Gossypol is a natural phenol derived from the cotton plant (genus Gossypium). Gossypol is a phenolic aldehyde that permeates cells and acts as an inhibitor for several dehydrogenase enzymes. It is a yellow pigment.

Among other things, it has been tested as a male oral contraceptive in China. In addition to its putative contraceptive properties, gossypol has also long been known to possess antimalarial properties.

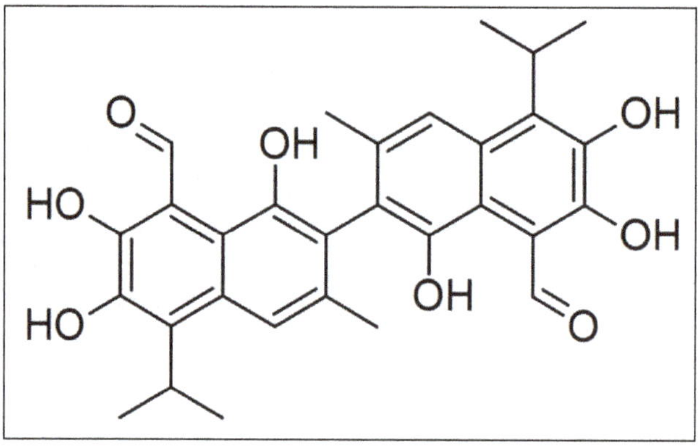

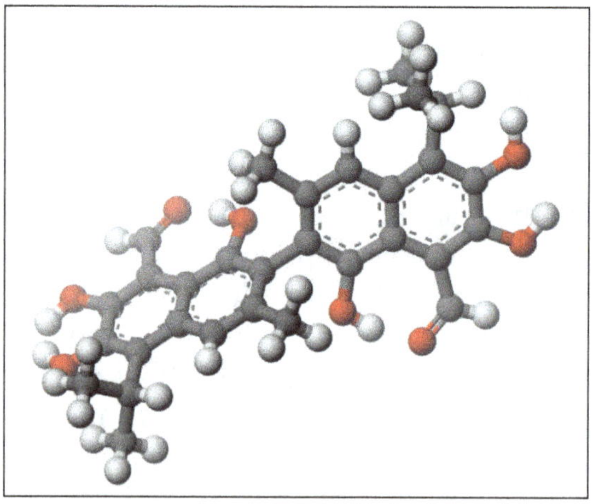

Biosynthesis

Gossypol is a terpenoid aldehyde which is formed metabolically through acetate via the isoprenoid pathway. The sesquiterpene dimer undergoes a radical coupling reaction to form gossypol. The biosynthesis begins when geranyl pyrophosphate (GPP) and isopentenyl pyrophosphate (IPP) are combined to make the sesquiterpene precursor farnesyl diphosphate (FPP). The cadinyl cation (1) is oxidized to 2 by (+)-δ-cadinene synthase. The (+)-δ-cadinene (2) is involved in making the basic aromatic sesquiterpene unit, homigossypol, by oxidation, which generates the 3 (8-hydroxy-δ-cadinene) with the help of (+)-δ-cadinene 8-hyroxylase. Compound 3 goes through various oxidative processes to make 4 (deoxyhemigossypol), which is oxidized by one electron into hemigossypol (5, 6, 7) and then undergoes a phenolic oxidative coupling, ortho to the phenol groups, to form gossypol (8). The coupling is catalyzed by a hydrogen peroxide-dependent peroxidase enzyme, which results in the final product.

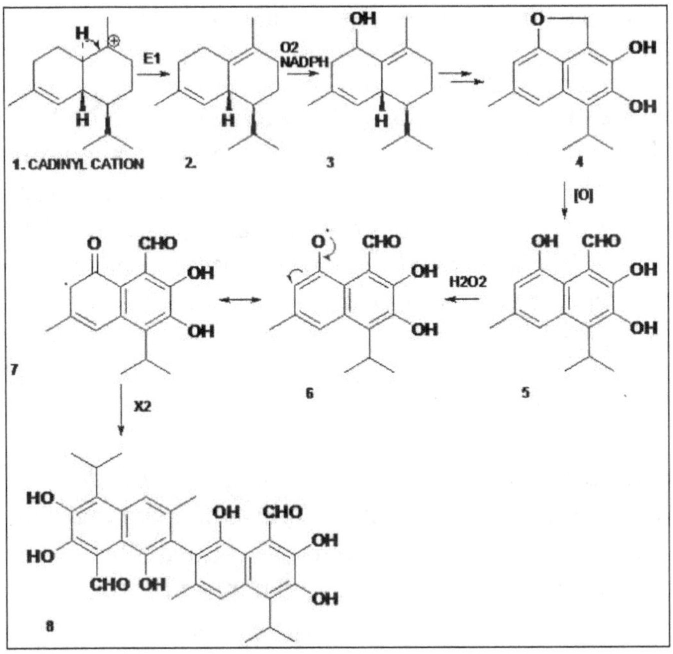

Contraception

A 1929 investigation in Jiangxi showed correlation between low fertility in males and use of crude cottonseed oil for cooking. The compound causing the contraceptive effect was determined to be gossypol.

In the 1970s, the Chinese government began researching the use of gossypol as a contraceptive. Their studies involved over 10,000 subjects, and continued for over a decade. They concluded gossypol provided reliable contraception, could be taken orally as a tablet, and did not upset men's balance of hormones.

However, gossypol also had serious flaws. The studies also discovered an abnormally high rate of hypokalemia among subjects. Hypokalemia—low blood potassium levels—causes symptoms of fatigue, muscle weakness, and at its most extreme, paralysis. In addition, about 7% of subjects reported effects on their digestive systems, and about 12% had increased fatigue. Most subjects recovered after stopping treatment and taking potassium supplements. The same study showed taking potassium supplements during gossypol treatment did not prevent hypokalemia in primates. The potassium deficiency may also be a result of the Chinese diet or genetic predisposition.

In the mid-1990s, the Brazilian pharmaceutical company Hebron announced plans to market a low-dose gossypol pill called Nofertil, but the pill never came to market. Its release was indefinitely postponed due to unacceptably high rates of permanent infertility. 5% to 25% of the men remained azoospermic up to a year after stopping treatment.

Researchers have suggested gossypol might make a good noninvasive alternative to surgical vasectomy.

In 1998, the World Health Organization's Research Group on Methods for the Regulation of Male Fertility recommended the research should be abandoned. In addition to the other side effects, the WHO researchers were concerned about gossypol's toxicity: the LD_{50} in primates is less than 10 times the contraceptive dose. This report effectively ended further studies of gossypol as a temporary contraceptive, but research into using it as an alternative to vasectomy continues in Austria, Brazil, Chile, China, the Dominican Republic, and Nigeria.

Toxicity

Food and animal agricultural industries must manage cotton-derivative product levels to avoid toxicity. For example, only ruminant microflora can digest gossypol, and then only to a certain level, and cottonseed oil must be refined. Genetically engineered cotton plants that contain little gossypol in the seed may still contain the compound in the stems and leaves.

Terpenes and Terpenoids

Natural products are the compounds which isolate from different natural sources such as plants, animals, microbes, insects, plant pathogens, and endophytes and marine. These are known as secondary metabolites since they are formed due to the enzymatic resections of primary metabolites (amino acids, sugars, vitamins, etc). Terpenes belong to the biggest class of secondary metabolites and basically consist of five carbon isoprene units which are assembled to each other (many isoprene units) by thousands of ways. Terpenes are simple hydrocarbons, while terpenoids are modified class of terpenes with different functional groups and oxidized methyl group moved or removed at various positions. Terpenoids are divided into monoterpenes, sesquiterpenes, diterpenes, sesterpenes, and triterpenes depending on its carbon units. Most of the terpenoids with the variation in their structures are biologically active and are used worldwide for the treatment of many diseases. Many terpenoids inhibited different human cancer cells and are used as anticancer drugs such as Taxol and its derivatives. Many flavorings and nice fragrances are consisting on terpenes because of its nice aroma.

Terpenes and its derivatives are used as antimalarial drugs such as artemisinin and related compounds. Meanwhile, terpenoids play a diverse role in the field of foods, drugs, cosmetics, hormones, vitamins, and so on.

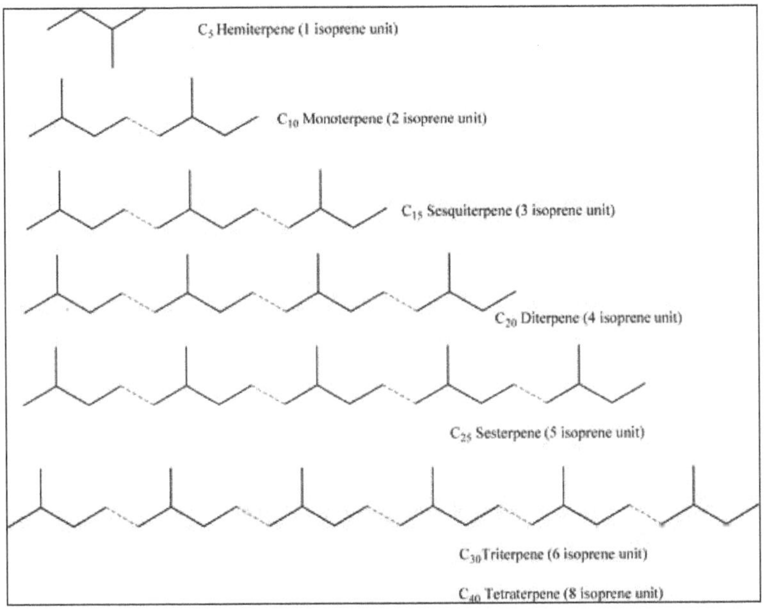

Classification of terpenes.

Monoterpenes

Monoterpenes consist of 10 carbon atoms with two isoprene units and molecular formula $C_{10}H_{16}$. These are naturally present in the essential and fixed oils of plants and related sources. Monoterpenes are structurally divided into the acyclic, monocyclic, and bicyclic type of compound. The compounds belong to this class usually have strong aroma and odor and are used in many pharmaceutical companies. Mixture of different monoterpene-based oils is used as fragrances for making perfumes and in other cosmetics. Most of the monoterpenes are active biologically with strong antibacterial activities. Several studies have shown in vitro and in vivo antitumor activity of many essential oils obtained from plants. The antitumor activity of essential oils of many species has been related to the presence of monoterpenes in their composition. Herein, we are discussing some of the recently published active monoterpenes.

Sesquiterpenes

Sesquiterpenes are the class of secondary metabolites consisting of three isoprene units ($C_{15}H_{24}$) and found in linear, cyclic, bicyclic, and tricyclic forms. Sesquiterpenes are also found in the form of lactone ring. Many of the latex in latex-producing plants contain sesquiterpene, and these are potent antimicrobial and anti-insecticidal agent. Artemisinin, a sesquiterpene lactone, one of the most active compounds in Artemisia annua shoots and roots.

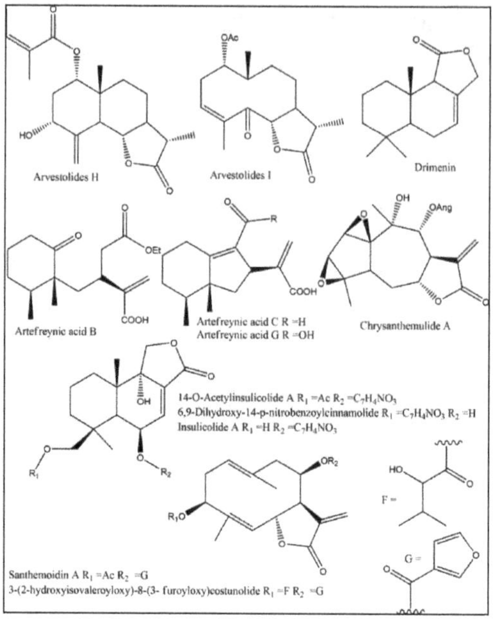

Structures of sesquiterpenes.

Diterpenes

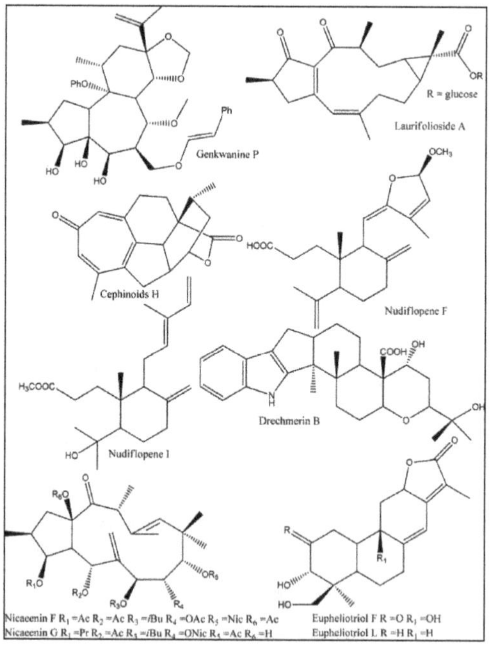

Structure of diterpenes.

Diterpenoids belong to a versatile class of chemical constituents found in different natural sources having $C_{20}H_{32}$ molecular formula and four isoprene units. This class of compounds showed significant biological activities including anti-inflammatory, antimicrobial, anticancer, and antifungal activities. Some of the diterpenes also have cardiovascular activity, such as grayanotoxin, forskolin, eleganolone, marrubenol, and 14-deoxyandrographolide. Kaurane and pimarane-type diterpenes are also biologically active metabolites isolated from the roots and leaves of different plants.

Sesterpenes

Sesterpenes consist of 25 carbon atoms with 5 isoprene units and molecular formula $C_{25}H_{40}$. These are naturally present in the fungus, marine organism, insects, sponges, lichens, and protective waxes of insects. These types of compounds are biologically active having anti-inflammatory, anticancer, antimicrobial, and antifungal activities.

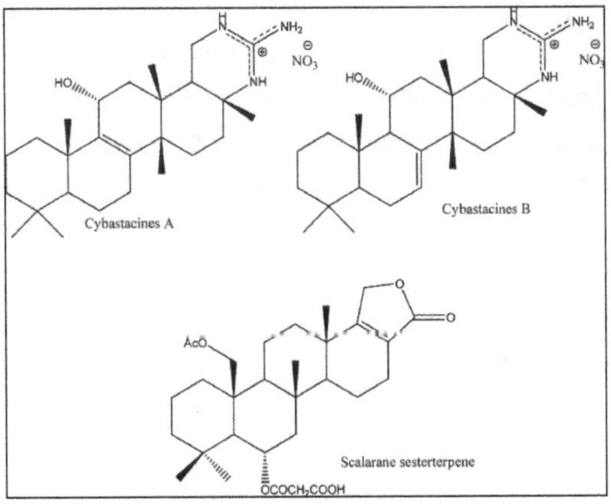

Structures of sesterpenes.

Triterpenes

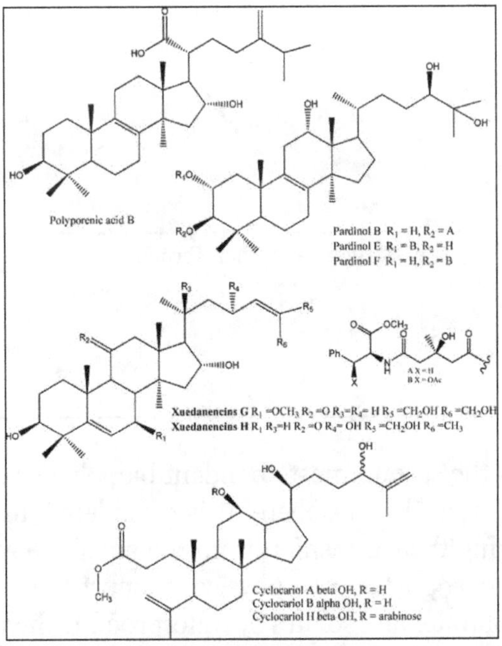

Structure of triterpenes.

A major class of secondary metabolites are known as triterpenes and it usually contains 30 carbon atoms consisting of 6 isoprene units. It is derived from the squalene biosynthetic pathway. Triterpenes have many methyl groups and it can be oxidized into alcohols, aldehydes, and carboxylic acids, which make it complex and differentiate it biologically. Triterpenes have many active sites for the glycosylation which converts it into another big class of compounds, namely, saponins (triterpene glycoside).

Meroterpenes

Meroterpenes are the secondary metabolites with partial terpenoid skeleton. Meroterpenoids were partially derived from mevalonic acid pathways and widely derived from animals, plants, bacteria, and fungi. Meroterpene biosynthesis expands the diversity available to isoprenoid pathways alone and allows for the assembly of natural products with highly unique structural attributes. Organisms belonging to the fungal kingdom have become proficient at exploiting this broad chemical synthesis platform for complex metabolite production.

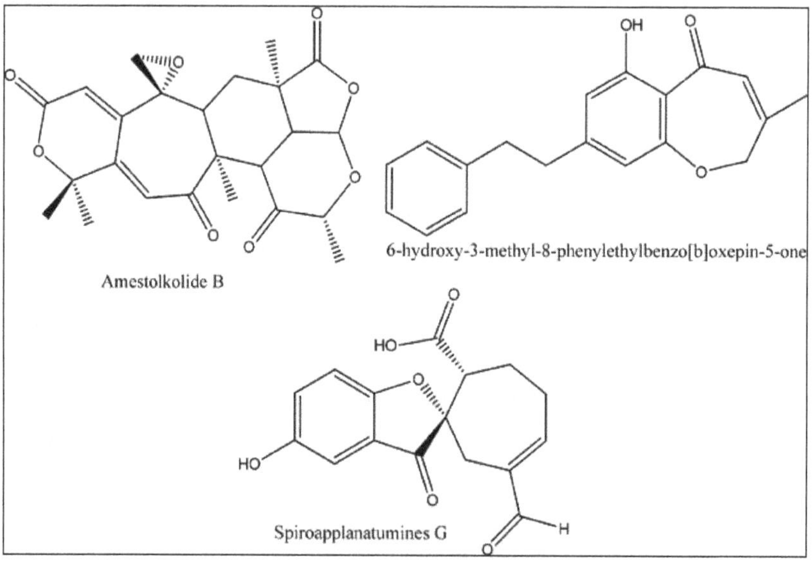

Structures of meroterpenes.

LIGNINS

Lignin, after cellulose, is the second most abundant biopolymer on Earth, accounting for 30% of the organic carbon in the biosphere. It is considered an important evolutionary adaptation of plants during their transition from the aquatic environment to land, since it bestowed the early tracheophytes with physical support to stand upright and enabled long-distance transport of water and solutes by waterproofing the vascular tissue. Although essential for plant growth and development, lignin is the major plant cell wall component responsible for biomass recalcitrance to industrial processing. The fact that lignin is a non-linear aromatic polymer built with chemically diverse and poorly reactive linkages and a variety of monomer units precludes the ability of any single enzyme to properly recognize and degrade it. Consequently, the use of lignocellulosic feedstock as a renewable and sustainable resource for the production of biofuels and bio-based materials will depend on the identification and characterization of the factors that determine plant biomass recalcitrance, especially the highly complex phenolic polymer lignin.

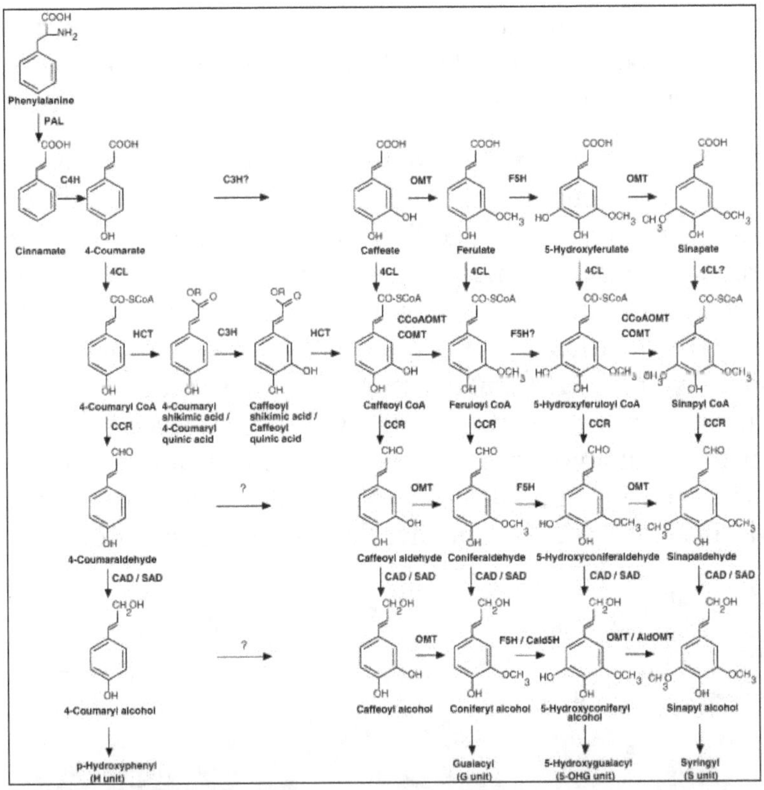

Around 450 million years ago, during the Late Ordovician, the first plant species started to occupy a challenging new niche, the terrestrial environment. As a consequence, pioneering land plants had to cope with several major stresses, including lack of structural support, harmful UV-B radiation, desiccation stress and potential co-evolving pathogens and herbivores. Among the physiological adaptations that arose during the evolution of early land plants, the emergence of the phenylpropanoid pathway was probably the most crucial evolutionary step. The ability to accumulate simple phenylpropanoids with absorbance maxima in the UV-B range allowed early land plants to resist UV irradiation and, consequently, made survival on land possible. Phenylpropanoids play a role in all aspects of plant responses towards abiotic and biotic stimuli, from mediating plant resistance towards pests to promoting a successful reproduction. Although the initial move of early plants onto land was facilitated by the acquisition of the phenylpropanoid metabolism, the true dominance of the terrestrial ecosystem by plants was made possible only with the development of cell walls reinforced with the phenylpropanoid polymer lignin, which bestowed the early tracheophytes with physical support to stand upright and enabled long-distance transport of water and solutes by waterproofing the vascular tissue.

In addition, the ability of lignin to resist degradation due to its chemically complex structure made this polymer an effective defensive barrier against herbivores and pathogens. Indeed, since plants do not possess a mechanism to degrade lignin, the deposition of lignin represents a non-recoverable investment of carbon and energy and, therefore, might be tightly regulated to guarantee that carbon skeletons requirement does not outpace availability.

Metabolic pathway for the biosynthesis of the three monolignols: p-coumaryl, coniferyl, and sinapyl alcohols. After produced in the cytoplasm, monolignols are transported to the cell wall where they will be oxidized by peroxidases and/or laccases prior to their incorporation into the polymer. Double squares represent the cell wall environment and the dark grey box indicates lignin units incorporated into the polymer: H (p-hydroxyphenyl), S (syringyl) and G (guaiacyl). Gene abbreviations: PAL (phenylalanine ammonia-lyase); C4H (cinnamate 4-hydroxylase); 4CL (4-coumarate-CoA ligase); C3H (p-coumarate 3-hydroxylase); HCT (p-hydroxycinnamoyl-CoA: quinate shikimate p-hydroxycinnamoyltransferase); CCoAOMT (caffeoyl-CoA O-methyltransferase); CCR (cinnamoyl-CoA reductase); F5H (ferulate 5-hydroxylase); COMT (caffeic acid O-methyltransferase); CAD (cinnamyl alcohol dehydrogenase).

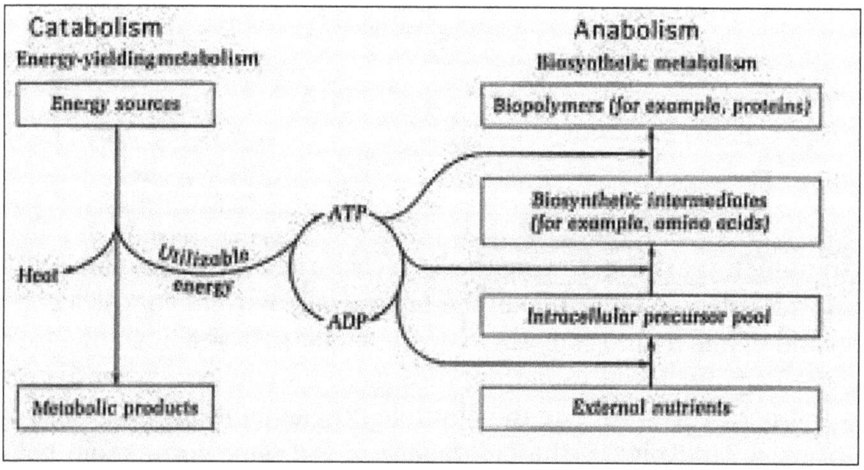

Lignin is a complex racemic heteropolymer produced by the oxidative combinatorial coupling of mainly three p-hydroxycinnamoyl alcohol monomers, the so-called monolignols. The shikimate pathway provides the aromatic amino acid phenylalanine that constitutes the entry point of the monolignol biosynthesis. After deamination of phenylalanine by the enzyme phenylalanine ammonia lyase (PAL), aromatic-ring modifications through hydroxylation and methylation followed by the transformation of the carboxylic moiety of the propane tail via esterification and reduction ultimately produce the three p-hydroxycinnamoyl alcohols, p-coumaryl, coniferyl and sinapyl alcohols, that differ in their degree of methoxylation.

Lignin biosynthesis, polymerization and distribution. Detailed structure of the plant cell wall shows the complex organization of the main components, in which the cellulose microfi brils are embedded in a matrix of hemicelulose, pectin and lignin. The spatio-temporal deposition of lignin during plant development is also shown.

When incorporated into the polymer, these monomers are called p-hydroxyphenyl (H), guaiacyl (G) and syringyl (S) units, respectively, and their individual contribution to lignin composition varies significantly among cell types, taxa and between tissues in the same plant. Phenylpropanoids other than the three canonical monolignols can also be incorporated into the polymer at varying levels, including hydroxycinnamates, hydroxycinnamyl acetates and hydroxycinnamyl aldehydes. This biopolymer is mainly deposited in the secondarily thickened cell walls of tracheary elements and fibers and, together with hemicellulose, forms a complex matrix in which cellulose microfibrils are embedded.

The deposition of lignin not only follows a developmental program but can also be triggered by several biotic and abiotic stresses.

In general, the lignin of angiosperm dicotyledonous plants is composed of G and S units and traces of H units, whereas the lignin from non-flowering vascular plants (i.e. lycophytes, ferns and gymnosperms) is mostly composed of G units with minor amounts of H units. Monocotyledonous lignin contains similar levels of both S and G units and the amount of H-units is relatively higher. Additionally, grass lignin contains significant amounts of the hydroxycinnamates ferulic acid and p-coumaric acid. Non-vascular organisms do not synthesize lignin, although bryophytes can accumulate soluble phenylpropanoids like flavonoids and lignans. Despite this proposed evolutionary scenario, notable exceptions were found in recent years including the discovery of lignin in peripheral cells of the red algae Calliarthron cheilosporioides M., the presence of lignin in non-vascular tissues of the liverwort Marchantia polymorpha L. and the recruitment of sinapyl alcohol for S lignin biosynthesis in suspension cell cultures from the basal gymnosperm Ginkgo biloba L.

Cell wall lignification, as well as the incorporation of each monomer, is regulated in a spatio-temporal fashion and varies between primary and secondary cell walls and among tissues. Lignin content increases with plant maturity. The incorporation of H and G units starts at the onset of lignifi cation, while only a few S units are detected during early stages of lignin deposition. Subsequently, coniferyl alcohol and increasing amounts of sinapyl alcohol are incorporated to form a mix of G and S units during secondary wall formation. In grass cell walls, hydroxycinnamates are also incorporated during secondary cell wall development and lignification. While ferulic acid is the major hydroxycinnamic derivative in young grass cell walls, p-coumaric acid is an indicator of cell wall maturity, since it is mainly esterified to side chains of S units and its incorporation follows the same deposition pattern of syringyl units. Differential distribution of lignin monomers is also observed in case of specific cell types. In most dicotyledonous plants, thickening of secondary walls occurs mainly in water-conducing cells of the xylem, which tend to contain higher levels of G units, and in structural fibers, which are typically enriched in S units. In grasses, parenchyma and other tissues such as epidermis and hypodermis present limited but significant lignification and xylary tissues and sclerenchyma accumulate high amounts of lignin and are highly enriched in S lignins. The ratio of S/G subunits in lignin also predicts the degree and nature of polymeric cross-linking. G-rich tissues have a more highly cross-linked lignin featuring a greater proportion of biphenyl and other carbon-carbon bonds, whereas S-rich lignin is less condensed, linked through more labile ether bonds at the 4-hydroxyl position. Additional methoxy groups on a lignin monomer results in less possible combinations during polymerization due to reduced available reactive sites. Consequently, S-rich lignin is more easily depolymerized than lignin rich in G units.

Monolignol Transport and Polymerization

Monolignols are synthesized in the cytoplasm and transported to the cell wall, where they are oxidized prior to their incorporation into the polymer. In contrast to the extensive knowledge of monolignol biosynthesis, little is known about the precise molecular mechanism of monolignol transport towards the cell wall. At least three systems were proposed: 1) exocytosis through ER-Golgi network; 2) passive diffusion through plasma membrane; and 3) active transport via specific membrane-localized transporters. Nevertheless, recent biochemical and genetic data strongly support the third hypothesis. By using isolated plasma and vacuolar membrane vesicles from Arabidopsis thaliana and different transporter inhibitors, Miao & Liu revealed several lines of evidence of the involvement of ABC-like transporters in monolignols transport across the plasmalemma and their sequestration into the vacuoles: 1) both processes were largely dependent on the presence ATP, which suggest that they are active processes; 2) uptake activity of both vacuolar and plasma membrane vesicles was strongly reduced by specific ABC transporter inhibitors in the presence of ATP; 3) the uptake process displays a typical protein-ligand binding kinetics, excluding the possibility of a passive diffusion process; and 4) vacuoles transporters presented high selectivity for monolignol glucoconjugates, whereas plasma membrane transporters preferentially transport monolignol aglycones. In another study, a role for an ATP-binding cassette transporter in p-coumaryl alcohol transport in Arabidopsis thaliana was demonstrated. Yeast cells expressing AtABCG29 exhibited increased tolerance to p-coumaryl alcohol while Arabidopsis loss-of-function mutants contained less lignin subunits and were more sensitive to p-coumaryl alcohol. This was the first identification of a monolignol transporter, although the characterization of specific transporters for coniferyl and sinapyl alcohol awaits further investigation.

After exported to the cell wall, monolignols are oxidized by peroxidases and/or laccases and further undergo polymerization through combinatorial radical-radical coupling. Both types of enzymes are present as large multigene families with redundant members, which make the study of lignin polymerization process difficult since generation of transgenic plants often results in no visible phenotype. In addition, their broad in vitro substrate specificity cannot offer precise information about their real substrates in plant Nevertheless, the involvement of peroxidases in lignin polymerization has been clearly demonstrated by a small number of studies reporting the generation of transgenic plants in which lignification was affected by the up- or down-regulation of a particular peroxidase gene. By contrast, only recently the role of laccases in lignification of stems was unambiguously demonstrated by reverse genetics. Berthet et al. demonstrated that disruption of AtLAC4 and AtLAC17 resulted in tissue-specific alterations to lignification in Arabidopsis thaliana stems. Still, genetic evidence for the functions of individual peroxidase or laccase genes in plants is mostly lacking, especially for monocotyledonous plants.

Following the oxidation, the resulting electron-delocalized monolignol radical presents unpaired electron density at the positions 1-, 3-, O-4-, 5- and 8- and, as the radical coupling is favored at the 8-position, coupling with another monolignol radical results in a mixture of dehydrodimers 8-5-, 8-8- and 8-O-4-linkages. After dimerization, the polymerization process will continue by the coupling of O-4-position of the dimer's phenolic end to the 8-position of the incoming monolignol radical. Lignin is further produced by the addition of monolignols to the growing polymer and not by the concatenation of preformed oligolignols. Because lignification is a strictly chemical process, any phenol localized in the lignification site of the cell wall can be used in the combinatorial radical coupling process, influenced by simple physical parameters such as temperature, pH, monolignol supply, concentration of hydrogen peroxide and oxidative enzymes and interactions with polysaccharides present in the matrix. Recently, a new hypothesis suggesting a strong biological control over the phenoxy radical coupling by the so-called dirigent proteins was proposed. However, current knowledge based on genetic data fall clearly in line with the chemical combinatory model.

Lignin and the Recalcitrance of Plant Biomass

Currently, the major bioenergy products are bioethanol, biodiesel and biogas. The bioethanol is produced through the fermentation of soluble sugars or starch, predominantly from sugarcane and maize kernel, and is often referred to as first-generation bioethanol. An emerging effort has been expended to increase the use of plant lignocellulosic biomass for the production of bioethanol, the so-called second generation bioethanol. Lignocellulosic biomass is composed of the polymers cellulose, hemicelluloses and lignin and is a renewable resource for the production of biofuels and biobased materials. However, lignocellulose presents a highly complex nature due to its molecular structure and heterogeneity, in which cellulose microfibrils are embedded in a matrix of hemicellulosic polysaccharides covalently cross-linked with the heterogenic and complex lignin. The recalcitrant nature of this mixture, mainly caused by the ability of lignin to resist enzymatic degradation, is one of the major obstacles to convert plant cell wall polysaccharides into fermentable sugars for biofuels production. The fact that lignin is a non-linear polymer built with chemically diverse and poorly reactive linkages and a variety of monomer units, makes this phenolic polymer the major plant cell wall component responsible for biomass recalcitrance. Consequently, the conversion of biomass to biofuel requires costly and harsh pre-treatments to degrade lignin and further allow the access to polysaccharides for saccharification.

Metabolic Engineering of Lignin to Modify Biomass Recalcitrance

An emergent strategy in the biofuel field is the genetic engineering of plants that either accumulate less lignin or produce lignins that are more amenable to chemical degradation, in an attempt to facilitate the processing of plant biomass. A common approach is to use genetic engineering to modify the expression of the enzymes responsible for lignin biosynthesis. The independent down-regulation of genes encoding seven enzymes of the monolignol pathway in alfalfa resulted in a reduction of the total flux into lignin, with more drastic effects found in the case of earlier enzymes in the pathway. In addition, striking differences in the lignin composition was observed dependent upon the targeted gene. While the down-regulation of C3H or HCT significantly increased the proportion of H units, the reduction in F5H or COMT expression resulted in lower levels of S units, consistent with the position of these enzymes in the pathway. A subsequent study in which the same transgenic lines were analyzed showed that biomass recalcitrance to both acid pre-treatment and enzymatic digestion is directly proportional to lignin content. Noteworthy, the saccharification efficiency of untreated biomass of the HCT and C3H lines, which had the lowest lignin content at < 50% of the wild type, was greater than that of pretreated wild type biomass. However, reducing lignin content to improve polysaccharide accessibility is not the only strategy, since the modification of its monomeric composition can also modify lignocellulose recalcitrance. Down-regulation of CAD in lignifying tissues in alfalfa did not result in any reduction of lignin content but showed an improvement in digestibility and pulping process. Poplar plants overexpressing F5H showed high levels of S-lignin and therefore digest and pulp more easily.

Since drastic reductions in lignin levels have negative effects on plant growth and development but even large shifts in lignin composition are still tolerable, the incorporation of atypical phenolic monomers into lignin through genetic engineering to produce a more easily degradable lignin is a new trend in the field. A remarkable example of such strategy comes from the concomitant down-regulation of COMT and up-regulation of F5H in Arabidopsis. Both enzymes are required to redirect the flux from coniferyl alcohol precursors towards sinapyl alcohol; the hydroxylation of coniferaldehyde by F5H produces 5-hydroxyconiferaldehyde that in turn is methylated by COMT. Since the hydroxylation step performed by F5H is essentially irreversible and 5-hydroxyconiferaldehyde cannot be converted to sinapaldehyde due to COMT down-regulation, CAD catalyzes the convertion of 5-hydroxyconiferaldehyde to 5-hydroxyconiferalcohol. This novel monomer behaves identically as a canonical monolignol, being exported to the cell wall and cross-coupling with normal lignin units to produce a polymer composed almost entirely from units that are undetectable in wild type plants. Therefore, the incorporation of "monolignol-substitutes" that can introduce readily cleavable linkages into the complex structure of lignin polymer seems feasible and can be a convenient option to achieve a more effective deconstruction of the plant cell wall.

Sugarcane as Bioenergy Crop

Sugarcane ethanol program in Brazil is the most successful example of efficient use of renewable energy. Brazil is the world's largest sugarcane producer and the leading country in ethanol production. In the 2009/2010 crop season approximately 612 million tons of sugarcane were produced, from which 25 billion liters of ethanol were generated. Sugarcane is a prominent bioenergy crop due to the ability to accumulate up to 50-60% of dry weight of stem as sucrose in the mature tissues, a unique feature within Poaceae. Although the production of sugar-based ethanol represents the most convenient option for sustainable energy in the case of sugarcane, lignocellulosic biomass can also be used, especially because sugarcane is among the most efficient biomass producers known. Sugarcane bagasse, the residue produced after sucrose extraction, is a plentiful lignocellulosic feedstock, composed by 39% cellulose, 25% hemicelluloses and 23% lignin, among other minor components. At present, the recovered bagasse is used for generation of heat and power for sugar processing into ethanol and also in electricity production that is sold into the grid. Therefore, the sugarcane industry could benefit from the production of second generation ethanol since the bagasse is promptly available after sucrose extraction and cellulosic ethanol could be co-produced and share part of the infrastructure used for the production of first generation ethanol. Most of our current knowledge of lignin metabolism is derived from studies in dicotyledonous herbaceous plants like Arabidopsis and alfalfa, but the mechanisms underlying lignin biosynthesis, polymerization and regulation are not necessarily conserved among all vascular plants. In addition, bioenergy crops may ultimately include dicotyledonous woody plants, such as eucalyptus and hybrid poplar, and monocotyledonous grasses, such as sugarcane and switchgrass. For instance, genetic information on lignin biosynthesis in sugarcane is scarce and limited to few studies reporting general expression profiling in which some phenyl propanoid genes were differentially expressed during sugarcane stem development or in relation to sucrose accumulation. Only recently, the first attempts to perform functional characterization of lignin-related genes in sugarcane were reported. The down-regulation of a sugarcane COMT gene reduced lignin content by up to 14%, resulting in an increase of 29% in saccharification yield.

In addition, evidence for the involvement of specific peroxidase isoforms in the lignification process were obtained by analyzing the activity, proteomic profile and expression pattern of class III peroxidases in sugarcane stems and suspension cell cultures. However, despite these recent advances, there is still much to be explored and determined in terms of gene expression and pathway regulation of lignin biosynthesis in sugarcane.

References

- Plant-metabolism, biochemistry-genetics-and-molecular-biology: sciencedirect.com, Retrieved 18 June, 2020
- Pathways-and-cycles, plant: britannica.com, Retrieved 16 April, 2020
- Yildiz F (2005). Phytoestrogens in Functional Foods. Taylor & Francis Ltd. pp. 3–5, 210–211. ISBN 978-1-57444-508-4
- Plant-Cell-Biochemistry-And-Metabolism2-Carbohydrate-Metabolism: plantcellbiology.masters.grkraj.org, Retrieved 17 January, 2020
- Flavonoids-in-agriculture, flavonoids-from-biosynthesis-to-human-health: intechopen.com, Retrieved 06 March, 2020
- An-overview-of-lignin-metabolism-and-its-effect-on-biomass-recalcitrance-261993156: researchgate.net, Retrieved 25 July, 2020
- Eisenhut M, Ruth W, Haimovich M, Bauwe H, Kaplan A, Hagemann M (November 2008). "The photorespiratory glycolate metabolism is essential for cyanobacteria and might have been conveyed endosymbiontically to plants". Proceedings of the National Academy of Sciences of the United States of America. 105 (44): 17199–204. doi:10.1073/pnas.0807043105. PMC 2579401. PMID 18957552

4

Agriculture Genomics

The usage of genomics and sequencing in agriculture for improving the crop productivity and livestock production is termed as agriculture genomics. Genetic engineering, reverse genetics, somatic fusion, genetically modified crops and plants, etc. are a few of its concepts. The topics elaborated in this chapter will help in gaining a better perspective of agriculture genomics.

Agricultural genomics is a rich field that has been contributing to advances in crop development for decades. From sequencing reference genomes to genotyping for genome-wide association studies to genomic prediction, advances in technology and applications have led to breakthroughs in crop improvement. These innovations have resulted in elite cultivars that have been selected for agriculturally desirable traits, including high yield, stress tolerance and pest resistance. However, the environments in which crops are grown are increasingly susceptible to changing conditions in the form of unstable weather patterns or ecologies. How do we best utilize genomic technologies to adapt to this rapidly changing world?

Notably, one potential way for genomics to lend itself to crop improvement and food security is through the collection-wide sequencing and classification of established seed banks or genebanks, in which important agricultural species are stored and maintained in large collections organized by taxonomy and origin. These rich resources reflect an amazing diversity of untapped genetic potential, but there are logistical challenges to extracting the maximum information from these records. In this issue of Nature Genetics, Stein and colleagues from IPK Gatersleben use genotype-by-sequencing data from more than 22,000 barley accessions from a genebank in Germany to analyze the diversity, redundancy and relationships in this large sample. The authors show that these 'molecular passport' data can be used to perform genome-wide association study analysis in the well-powered dataset, thus providing insight into the selection patterns of desired traits. The large-scale genetic characterization of the entirety of this gene-bank collection provides a framework for a generalizable approach that can be applied to other crops as well.

The preservation of seed resources ensures that the natural genetic diversity captured by the collections will not be lost. Accurately cataloging these resources by using genomic data provides precise and usable information for breeders and scientists, while simplifying the effort by identifying redundancy in the collection. Expanding sequencing of other genebanks could eventually allow for a comparative and holistic view of the world-wide composition of agriculturally important crop collections, thereby providing raw material for breeding or engineering improved agricultural plant species.

The development of improved crops is a priority for scientists and policy makers alike. Indeed, one of the United Nations' Sustainable Development Goals, intended to ensure a better future for all, is Zero Hunger. With the twin pressures of population growth and climate change, new strategies are needed to cope with disruptions to food growth and security, and these policies must be practical, ethical and sustainable. Outside the realm of policy, innovation is likewise needed on the scientific front. Applying genomic approaches to existing resources will greatly expand the ability to meet future challenges associated with food security, increasing human resilience in the face of a changing world—a world that should be hunger free.

COMPARATIVE GENOMICS

Genomics information gained in one organism has applications in other even distantly related organisms. Comparative genomics allows one to achieve a greater understanding of evolution. Comparing genomes also reveals what is common and what is unique between different species at the genome level and helps to identify both coding and non-coding genes and regulatory elements.

How to use Genomics for the Betterment of Agriculture?

Molecular plant breeding aims to improve crop variety in context to its yield, quality and resistance by using latest innovations made in genetics and genomics. The understanding about the association between genotype and phenotype has been growing with the help of genomics tools. One of the major applications of genomics directly related to breeding is the identification of molecular markers associated with the trait of interest. Such markers have proven highly significant for the plant breeders in plants by overcoming the constraints of phenotypic selection, which sometimes is unable to identify individual with highest breeding value as shown below:

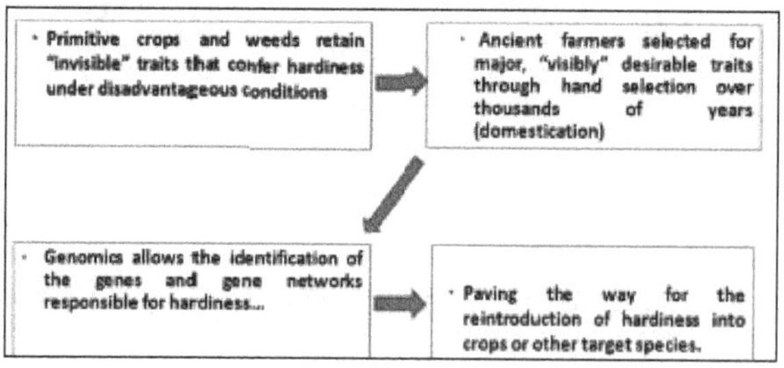

Uses of Different Tools of Genomics in Agriculture

Why Research on Genomics in Agriculture?

Human beings cannot live without plants. Directly or indirectly, most of the world's food comes from agriculture. Agriculture offers renewable sources of energy and materials.

Molecular characterization: exploring phylogeny and evolution Molecular characterization plays an important role in uncovering the history of organisms, i.e. their evolution from wild ancestors and to classify them into appropriate groups on the basis of genetic diversity, distinctiveness and population structure of organisms. This is especially important in cases where morphological markers have failed to give accurate results or even the results are misleading. For example, the rice lines Azucena and PR 304 have been classified as indicas using morphological characters whereas they behave like japonicas in crossing studies. These samples are, however, clearly revealed to be japonicas when analysed by molecular markers.

Quantitative Trait Loci (QTL) Mapping

The term was coined by Gelderman in 1975. It is defined as regions of genome associated with effect on quantitative trait which is controlled by number of genes having cumulative effect. The process of constructing linkage maps and conducting QTL analysis to identify genomic regions associated with traits is known as QTL mapping.

Principle: 'QTL mapping' is based on the principle that genes and markers segregate via chromosome recombination (called crossing-over) during meiosis, thus allowing their analysis in the progeny. Genes or markers that are close together or tightly linked will be transmitted to gether from parent to progeny more frequently than genes or markers that are located further apart. Markers that have a recombination frequency of 50% are described as 'unlinked' and assumed to be located far apart on the same chromosome or on different chromosomes.

Mapping population: The construction of a linkage map requires a segregating population (i.e. a population derived from sexual reproduction). The parents selected for the mapping population will differ for one or more traits of interest. Population sizes generally range from 50 to 250 individuals; however, for high-resolution mapping larger populations are required. Several different populations may be utilized for mapping as:

- F2 populations are derived by selfing of F1 hybrids and backcross (BC) populations are generated by crossing the F1 hybrid to one of the parents. These are the simplest types of mapping populations developed for self-pollinating species. Their main advantages are that they are easy to construct and require a short time to produce.

Choice of contrasting parents
(i.e. susceptible and resistant parents)

↓

Parental polymorphism
(Different molecular markers are used to carry out parental polymorphism and those which are polymorphic between parents are further used for genotyping of mapping population)

↓

Generation of mapping population

↓

Genotyping of mapping population
(Polymorphic markers selected from parental polymorphism are used to carry out genotyping of mapping population used)

↓

Phenotyping of mapping population
(Mapping population is phenotyped against the trait studied)

↓

The data generated from genotyping and phenotyping is used for the construction of QTL map with the help of appropriate software

Steps to be followed for QTL mapping.

- Recombinant inbred (RI) lines are constructed by inbreeding the individual F2 plants, resulting in a series of homozygous lines, each containing a unique combination of chromosomal segments from the original parents. The major disadvantage of RI population is the length of time needed to produce them, because usually six to eight generations are required.
- Doubled-haploid (DH) populations can be produced by regenerating plants by the induction of chromosome doubling from pollen grains. But the production of DH populations is only possible in species that are amenable to tissue culture (e.g. cereal species such as rice, barley and wheat). The major advantages of RI and DH populations are that they produce homozygous or true-breeding lines that can be multiplied and reproduced without genetic changes, representing them as eternal resources for QTL mapping.

Till date QTL mapping has been done in several crops, such as tomato, pea and chickpea.

Table: Methods used to detect QTL.

Methods for statistical analysis	Description
Single-marker analysis	• Also known as 'single-point analysis.' • This is the simplest method for detecting QTL associated with single markers. • Single-marker analysis is done using t tests and analysis of variance (ANOVA.
Simple interval mapping (SIM)	• In this method, linkage maps and intervals between adjacent pairs of linked markers along chromosomes are used for analysis. • Analysis is done using MapMaker/QTL and Qgene.

Composite interval mapping (CIM)	• In this method, interval mapping for a single QTL is combined with other marker associated with other QTL. • Analysis is performed using QTL Cartographer, MapManager QTX and PLABQTL.
Multiple interval mapping (MIM)	• It is an extension of interval mapping to multiple QTL. • Can allow interaction between QTL.

MAS

In MAS, plants are selected for one or more (up to 8–10) alleles. Major steps involved in MAS are:

- Identification of molecular marker(s) associated with trait(s) of interest.
- Validation of identified marker(s) in the genetic background of the targeted genotypes which are to be improved.

Advantages of MAS

The use of DNA markers for screening and selecting plants in a breeding programme provides several advantages and is therefore very attractive to plant breeders as:

- DNA marker studies can be done from almost any plant tissue. Plants can be screened at the seedling stage or even as seeds, thus, allowing early selection for traits which may be expressed in adult plants only (i.e. grain or fruit quality, male sterility and photoperiod sensitivity, etc).
- Target alleles that are difficult, expensive and time consuming to score phenotypically can be selected with the assistance of markers.
- Selections can be made on a single plant basis, which is not possible by phenotypic selection. Poor heritability is not a problem if selection is based on marker information.
- For traits with complex inheritance every individual genetic component contributing to the trait can be selected separately. Also, multiple characters that would normally be epistatic (i.e. they show a certain positive or negative effect only in combination with each other) can be maintained and ultimately fixed.
- Markers can also be applied in the choice of parents in crossing programmes. Here they can either help to maximize diversity and support the exploitation of heterosis or they can minimize diversity, if gene complexes built up in elite inbred germplasm are to be preserved.

A number of varieties have been released through MAS in the world in different crops like barley, bean, rice, soyabean, tomato and wheat. In India, five varieties named MAS 946-1 (drought-tolerant rice), Pusa 1460 or Improved Pusa Basmati 1 and RP BIO 226 (disease-resistant rice) Vivek quality protein maize (QPM) 9 (high QPM) and HHB 67-2 (disease-resistant pearl millet) have been released through MAS.

MABC

Through MABC one or more (up to 6–8) donor alleles are transferred to an elite line. The goal of backcrossing is to move a single trait of interest, such as drought-tolerance, high-productivity or disease resistance, from a donor parent to progeny. MABC enables to monitor the transmission of the trait gene via a genetically linked marker that can be easily screened, significantly accelerating backcrossing programmes and reducing the time to release of commercially important plant lines or breeding stock.

Once the markers associated with a trait of interest are identified through linkage mapping, association mapping (AM), AB-QTL or transcriptomics approach etc., the next step is to use these markers in the breeding programme. Thus, the MABC uses linked markers to select the target gene/QTL from the donor parent and the unlinked markers to recover recurrent parent (RP). Traditional approaches of recovery of RP genome take up to six BC generations but the use of markers enables to achieve the same even in BC_2.

In recent years, the MABC approach is becoming more popular in developing countries such as India, China, Bangladesh and Thailand where public sector in collaboration with the international agriculture research centres have been working towards developing the improved varieties in their targeted environments. Some successful examples of MABC are: common bean, wheat, maize and lettuce.

MARS

In the majority of traits of interest, quantitative variation is controlled by many QTL each with minor effect. Minor QTL show an inconsistent QTL effect in different environments and over different seasons. Even when the effect of these minor QTL is consistent, their introgression into the desired genotype through MABC becomes extremely difficult as a larger number of progenies are required to select appropriate lines. In such cases, MARS is useful for pyramiding superior alleles at different loci/QTL (up to 20–30) in a single genotype.

MARS is based on the identification of markers associated with trait and estimation of their effect. This approach involves multiple cycles of marker-based selection that includes: (i) identification of F2 progeny that contain favourable alleles for most QTL; (ii) recombination of the selected progenies to the selfed ones; and (iii) repetition of these two cycles. The use of MARS is larger in case where prior information of the QTL is available and the response decreases as the knowledge of the number of minor QTL associated with the trait of interest decreases. Recently, some international agricultural research centres in collaboration with Generation Challenge Programme (GCP) have also initiated MARS in crops such as chickpea, sorghum, rice, cowpea, etc. for pyramiding favourable drought-tolerant alleles.

Advanced Backcross (AB)-QTL Analysis

AB-QTL aims at simultaneous detection and transfer of useful QTL from the wild/un-adapted relatives to a popular cultivar for improvement of a trait. Thus, a superior cultivar/variety is crossed with a wild species leading to the production of a backcross population (BC_2, BC_3) and molecular markers are used to monitor the transfer of QTL by conventional backcrossing. The advanced backcross approach has been successfully utilized in different crops, including tomato, rice, wheat, maize and barley.

Gene Pyramiding

Gene pyramiding is a remarkable process of combining several favourable genes/alleles from different genotypes into a single elite/commercial cultivar. Gene pyramiding is often performed through MAS and is widely used for combining multiple disease resistance genes against different races of a pathogen. Pyramiding is also possible through conventional breeding but phenotypically testing individual plants for all traits is time-consuming and very difficult. The most frequent strategy of pyramiding is combining multiple resistance genes. Different resistance genes can be combined in order to develop broad-spectrum resistance to, e.g. diseases and insects. An example for the combination of resistance QTL is the pyramiding of a major stripe rust resistance gene and two QTL in the same genotype.

AM

AM is an alternative strategy to linkage mapping for identifying a marker associated with trait. It utilizes natural variation and linkage disequilibrium existing in natural population to identify the gene(s) associated with trait rather than the offspring of crosses and associations in natural populations are typically on a much finer because they reflect historical recombination events. Linkage disequilibrium is a non-random association of alleles at different loci, describing the condition with unequal (increased or reduced) frequency of the haplotypes in a population at random combination of alleles at different loci.

Based on the aim of study, AM uses one of following two approaches: (i) candidate-gene AM, which relates polymorphisms in selected candidate genes which control phenotypic variation for specific traits and (ii) genome-wide AM, or genome scanning, which surveys genetic variation in the whole genome to find out signals of association for various complex traits.

Advantages of AM over Linkage Mapping

- Potential for increased QTL resolution.
- Iincreased sampling of molecular variation.
- In general, conventional linkage analysis uses a mapping population such as

F2 lines, BC population and recombinant inbred lines (RILs), which is a commonly used method for trait mapping. However, such mapping populations are derived from a few cycles of recombination events, hence limit the resolution of genetic maps and localize QTL from 10 to 20 cM intervals only.

- Results from AM are more applicable to a much wider germplasm base.

REVERSE GENETICS

The number of sequenced genes whose function remains unknown has soared in recent years with the advent of affordable genome sequencing. Even in those species whose entire genome has not been sequenced, there are genes whose sequence is known but whose function is not yet well understood. Bioinformatics has been invaluable in investigating the function of these genetic components, and powerful in silico techniques have been developed for the analysis of genome sequence information, but the elucidation of gene function must always be verified in vivo using genetic analysis. Reverse genetics is a powerful tool that establishes a direct link between the biochemical function of a gene product and its role in vivo. Several approaches have been developed for plants, some of which are applicable to many species and each of which has advantages and limitations.

Virus-induced Gene Silencing

Virus-induced gene silencing (VIGS) has emerged as one of the most adaptable techniques in recent years but its effects are transient in nature. VIGS is performed by cloning a 200–1300 bp cDNA fragment from a plant gene of interest into a DNA copy of the genome of an RNA-virus and transfecting the plant with this construct using Agrobacterium. Double-stranded RNA from the viral genome, including the sequence from the gene of interest, is formed during viral replication. The double-stranded RNA molecules are degraded into small interfering RNA (siRNA) molecules by the plant Dicer-like enzymes, thus activating the siRNA silencing pathway. This results in the degradation of the target gene transcript leading to a knockout or knockdown phenotype for the gene of interest. The most common vectors currently used are based on the tobacco rattle virus (TRV) but, recently, the apple latent spherical virus (ALSV) has been reported to be even broader in its host range and to have minimal side effects. VIGS has been used in a wide variety of plants including both dicot and monocot species. The advantages of this technique compared to other reverse genetics techniques include the facts that it is quite rapid and inexpensive and, because the phenotype is transient, deleterious loss-of-function effects may sometimes be observed without concomitant lethality. In addition, the function of several homologous genes may be affected with a single construct. Further, it is a technique limited only by the host range of the virus used, making it applicable to many species. However, the phenotypes induced using this technique are not heritable and so cannot be used for genetic engineering. In addition, the silencing level is variable depending on the construct and the growth conditions used and it is rare that gene(s) are completely silenced. A final disadvantage of this technique is that it has not yet been developed for all plant species.

RNA-mediated Interference

RNA-mediated interference (RNAi) is similar in action to VIGS, but is heritable in nature and so offers a different scope of investigation. In this technique, a DNA construct that produces either single- or double-stranded RNA complementary to the gene of interest is introduced into a cell where it activates the RNA silencing pathway and degrades some or all of the transcripts from the gene of interest as. There are several techniques commonly used to activate the RNAi pathway in plants but the one growing fastest in popularity is the use of artificial microRNAs (amiRNA). In this procedure, a microRNA (miRNA) gene carrying a 21 bp insert complementary to the target gene is transformed into the plant where it is processed by the Dicer pathway to produce the small silencing RNA molecules that target endogenous transcript(s) for degradation. A modified version of this techniques involves the production of interfering RNAs using promoters that are temporally or spatially specific or that are inducible by some exogenous factor.

The advantages of RNAi-based reverse genetics include the facts that a partial loss of function can be achieved, silencing is directed against a specific gene(s) so the screening of large populations is not required, and transcripts of multiple genes from a family can be silenced by a single construct. This feature is especially useful in plants given that many plants are polyploid or have experienced partial polyploidisation at some stage in their evolution. Other advantages of this technique are that the induced phenotypes are dominant, they can be observed in the T1 generation, and stable inheritance of the transgenic RNAi gene makes the technique suitable for genetic engineering. A disadvantage of the RNAi approach is that some genes are resistant to silencing by exogenous RNA, probably because of sequence or structural features of these genes. In addition, transcripts of genes that are similar in sequence to the target locus may be down-regulated inadvertently as well as the transcripts from the actual target gene. Such 'off-target' silencing may be difficult to interpret. Like VIGS, the silencing level may also be variable and rarely are the genes completely silenced. completely silenced. Furthermore, because RNAi involves transgenic loci that can be silenced in future generations of transgenic lines, the stability of the RNAi effect can sometimes be problematic. This technique, or older versions of it, have, however, been used in many plants to improve productivity or increase disease resistance as well as for basic investigations of gene function.

Insertional Mutagenesis

T-DNA or transposon insertion has been exploited to create disruptions in target genes of interest, introduce new genes, or activate endogenous genes in the plant genome.

For insertional knockouts this technology has been used in both monocotyledonous and dicotyledonous plants and remains one of the most effective ways of performing reverse genetics in many species. A population of plants each having an insertion(s) at a unique site in the genome is generated either by transformation (T-DNA) or transposon activation. Such populations with genome-wide insertions have been generated for several plant species as a reverse genetic service. Plants carrying an insertion in a gene of interest can be identified by screening the population with PCR using one gene-specific primer and one insertion based primer. Gene disruption studies using this technique typically result in a total loss of function and the insertion can be easily followed, using PCR, in a population of plants where it is segregating. Some of the disadvantages include the facts that phenotypes may not be obvious if the gene function is redundant and insertions in essential genes will typically result in lethality making these types of genes difficult to examine using this technique. In addition, with T-DNA insertion libraries, any given plant will carry at most a few insertions, therefore, very large populations (hundreds of thousands of plants) are needed to achieve genome saturation. This problem is alleviated for species in which endogenous transposons can be activated, but for those that don't have this advantage, generating and screening very large populations for T-DNA insertions is difficult.

Fast Neuron Mutagenesis

Fast neutron bombardment is used to generate deletions and chromosome rearrangements of various sizes randomly in the genome. For reverse genetics the major fast neutron technique currently used in plants is Deletagene. In this technique, seeds are mutagenised with fast neutron radiation and deletions are identified using PCR on pooled DNA using primer sequences that flank the gene of interest. Amplification time is restricted so as to preferentially allow amplification of the mutant DNA where a smaller PCR product (deletion) is generated. The Deletagene system has been used very effectively in both Arabidopsis and rice but is laborious because of the number of plants that must be screened and has limitations in terms of the sizes of deletions that can be recovered. A modified PCR detection strategy involving more steps than the Deletagene protocol has recently been tried on a Medicago truncatula fast neutron population. This involves the use of two rounds of PCR using nested primers. In the first round of PCR amplification a 'poison primer' is included that has been targetted to a region of the amplicon that is deleted in the mutant. This poison primer causes aberrant amplification of a partial product from the wild-type but not the mutant DNA. In addition, the first PCR product is digested with a restriction endonuclease whose target cleavage site has been deleted in the mutant. Thus, during the second round of PCR, the mutant product should be better represented than it was initially, allowing detection of the mutation even in very large pools of samples. The authors have named this technique 'De-TILLING', although its connection to the original TILLING technique is unclear. Another alternative screening protocol that has been tested in rice and in clementine involves hybridizing genomic DNA to cDNA-based microarrays to detect fast neutron-induced deletions in these plants. Such array-based technology may be useful for this type of reverse genetics screen in other systems as well. For these projects, genomic DNA was labelled and hybridised to the oligonucleotide arrays in order to detect genes, or groups of genes, that had been deleted in the mutant lines.

Chemical Mutagenesis and Tilling

Chemical mutagenesis using ethylmethane sulphonate (EMS) or ethyl nitrosourea (ENU) induces point mutations in DNA in all species in which it has been tested. Generation of mutant populations using these or similar mutagens was an established practice long before the advent of gene sequencing and reverse genetics. Mutations induced using these mutagens are distributed in the genome randomly and because point mutations are less damaging than large rearrangements, a high degree of saturation can be achieved in a mutant population, more easily enabling the examination of gene function on a genomic level. In addition, while most reverse genetics techniques provide only loss-of-function alleles, chemical mutagenesis can result in either loss-of-function or gain-of-function mutations causing null, hypomorphic, neomorphic or hypermorphic phenotypes. While hypomorphic alleles may be most useful for determining wild-type gene function, neomorphic or hypermorphic alleles are more likely to be dominant and so are more likely to cause an observable phenotype in cases of either redundant or essential genes. The frequency of missense alleles is on average three times higher than that of non-sense alleles, but it is difficult to predict how many of these mutations will have an effect on gene function since many of them may not alter the gene product(s) significantly. Examples of dominant point mutations that do have an effect on gene function have, however, been well documented in several cases, for example, the ethylene response pathway, leaf polarity determination and host-pathogen defense.

For forward genetics, mutagenised populations are screened for overt phenotypes, and then genetic and molecular analysis is used to discern the sequence of the gene responsible. The difficulty with using point mutations for reverse genetics is that there have not been many direct, cost-effective ways of screening for individuals from the mutagenised population that carry mutations in specific genes of interest. The advent of TILLING introduced such a method of screening at a reasonable cost within a reasonable time frame. Most TILLING operations use the technique described by Colbert et al. that employs a mismatch-specific endonuclease found in celery for identifying SNPs in genes of interest. In this procedure, DNA from several different plants is pooled and then used as a template for PCR amplification with fluorescently tagged, gene-specific primers. The PCR products (amplicons) are denatured and allowed to randomly re-anneal before being digested with a celery juice extract (CJE). Mismatches in the amplified DNA occur, following renaturation, if a pool of DNA includes at least one plant with a mutation in the amplified region. When a DNA strand from a mutant amplicon anneals to the complementary strand of the wild-type amplicon a heteroduplex is formed that presents a target for the CJE enzyme. When cleavage of some of the PCR products at the mismatch occurs, the novel fragments can be detected using a LI-COR DNA

Analyser (LI-COR Biosciences, Lincoln, NE) or other equivalent DNA separation technology, thus identifying plants with mutations in the gene of interest. Modifications of this procedure, such as the use of an Arabidopsis endonuclease (ENDO-1) instead of CJE, have been tested with similar results. TILLING is robust, and has worked well in a wide variety of plant species. In addition, until recently, it has been the least expensive and one of the only techniques for identifying de novo SNPs or point mutations in genomes for which the entire sequence is not available. However, a mutagenised population arrayed for TILLING must be available and for many species the development of such a population is expensive and time consuming. Further, the technique itself is labour-intensive and it has proven more difficult to obtain large numbers of mutations in some genes because of the technical requirements for primer design and visualisation using CJE and the Li-cor gel apparatus. Finally, results must be verified by sequencing adding to the time and cost.

Next-generation Sequencing for Reverse Genetics

With the advent of next-generation sequencing (NGS) technology it is now conceivable to Sequence Candidate Amplicons in Multiple Parallel Reactions (SCAMPR) for mutations in any gene at costs comparable to current TILLING procedures. Direct sequencing identifies the position and nucleotide change for each mutation at the time of detection and also provides haplotype information (which is currently not available from TILLING data). The question of which NGS technology would be the simplest and most cost effective for screening populations by SCAMPRing still needs to be addressed. In addition, the pooling procedures needed for practical application of this technology to mutation/SNP detection must be determined. There are two major classes of NGS machines for which a reasonable amount of data is already available: short read (25–75 bp; for example the Illumina Genome Analyser or ABI SOLiD) and longer read (400–500 bp; like the Roche454). The 454 technology is more expensive than the short-read technologies at this time, but should require less depth of sequencing to distinguish between bona fide mutations and background sequencing errors within a population. Both 454 and Illumina technologies have now been tested on TILLING populations, but not directly compared with other TILLING techniques. Keygene NV has tested the Roche 454 GS FLX using 5-fold pooled DNA for TILLING. They sequenced DNA from 15 000 M2 tomato plants (representing 3008 M1 lines) in 28 pools on a full picotitre plate using a three-dimensional labelling strategy with 12X, 8Y and 8Z column, row and plate pools. They amplified a region of 287 bp from a single locus and obtained total of 580 471 sequencing reads which they analysed using in-house software designed for this purpose (not publicly available). On the basis of this trademarked protocol (named KeyPoint technology), they identified two mutations that were confirmed by Sanger sequencing.

The Comai lab at the University of California at Davis Genome Center has also been testing TILLING by NGS but using the Illumia GA2 machine to sequence PCR amplicons. They offer a TILLING service in both rice and tomato as well as workshops to train other laboratories in this novel technique. This group is using a two-dimensional pooling strategy of 96-fold row pools and 64-fold column pools. One of the most difficult problems with using NGS for mutation detection is separating the true mutations from the sequencing errors generated by chance in the thousands of reads that are produced in each run. The UC Davis TILLING lab reports a variable success rate for this that is dependent on the species and pooling depth, but averages <0.7.

GENETIC ENGINEERING FOR PLANT

The field of genetic engineering has developed rapidly due to the greater understanding of deoxyribonucleic acid (DNA) as the chemical double helix code from which genes are made. The term genetic engineering is used to describe the process by which the genetic makeup of an organism can be altered using "recombinant DNA technology." This involves the use of laboratory tools to insert, alter, or cut out pieces of DNA that contain one or more genes of interest.

Developing plant varieties expressing good agronomic characteristics is the ultimate goal of plant breeders. With conventional plant breeding, however, there is little or no guarantee of obtaining any particular gene combination from the millions of crosses generated. Undesirable genes can be transferred along with desirable genes; or, while one desirable gene is gained, another is lost because the genes of both parents are mixed together and re-assorted more or less randomly in the offspring. These problems limit the improvements that plant breeders can achieve.

In contrast, genetic engineering allows the direct transfer of one or just a few genes of interest, between either closely or distantly related organisms to obtain the desired agronomic trait. Not all genetic engineering techniques involve inserting DNA from other organisms. Plants may also be modified by removing or switching off their own particular genes.

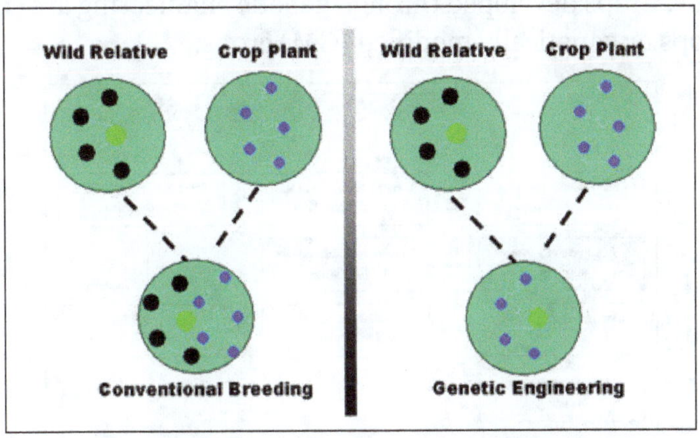

Comparing conventional breeding and genetic engineering.

Conventional Breeding	Genetic Engineering
• Limited to exchanges between the same or very closely related species. • Little or no guarantee of any particular gene combination from the million of crosses generated. • Undesirable genes can be transferred along with desirable genes. • Takes a long time to achieve desired results.	• Allows the direct transfer of one or just a few genes, between either closely or distantly related organisms. • Crop improvement can be achieved in a shorter time compared to conventional breeding. • Allows plants to be modified by removing or switching off particular genes.

Genes are molecules of DNA that code for distinct traits or characteristics. For instance, a particular gene sequence is responsible for the color of a flower or a plant's ability to fight a disease or thrive in extreme environment.

Nature's Own Genetic Engineer

The "sharing" of DNA among living forms is well documented as a natural phenomenon. For thousands of years, genes have moved from one organism to another. For example, Agrobacterium tumefaciens, a soil bacterium known as 'nature's own genetic engineer', has the natural ability to genetically engineer plants. It causes crown gall disease in a wide range of broad-leaved plants, such as apple, pear, peach, cherry, almond, raspberry, and roses. The disease gains its name from the large tumor-like swellings (galls) that typically occur at the crown of the plant, just above soil level. Basically, the bacterium transfers part of its DNA to the plant, and this DNA integrates into the plant's genome, causing the production of tumors and associated changes in plant metabolism.

Application of Genetic Engineering in Crop Production

Genetic engineering techniques are used only when all other techniques have been exhausted, i.e. when the trait to be introduced is not present in the germplasm of the crop; the trait is very difficult to improve by conventional breeding methods; and when it will take a very long time to introduce and/or improve such trait in the crop by conventional breeding methods. Crops developed through genetic engineering are commonly known as transgenic crops or genetically modified (GM) crops.

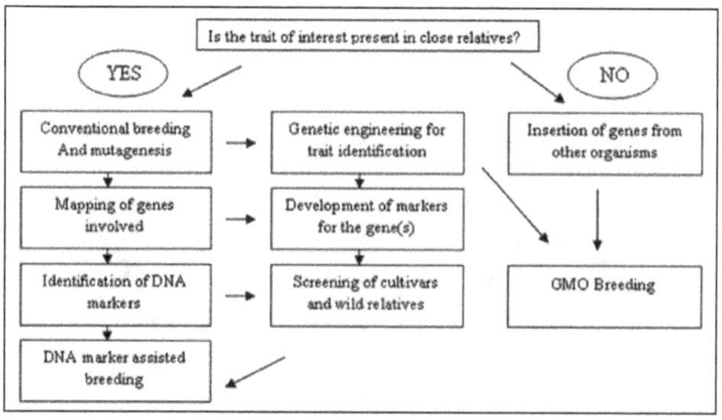

Modern Plant Breeding.

Modern plant breeding is a multi-disciplinary and coordinated process where a large number of tools and elements of conventional breeding techniques, bioinformatics, molecular genetics, molecular biology, and genetic engineering are utilized and integrated.

Development of Transgenic Crops

Although there are many diverse and complex techniques involved in genetic engineering, its basic principles are reasonably simple. There are five major steps in the development of a genetically engineered crop. But for every step, it is very important to know the biochemical and physiological mechanisms of action, regulation of gene expression, and safety of the gene and the gene product to be utilized. Even before a genetically engineered crop is made available for commercial use, it has to pass through rigorous safety and risk assessment procedures.

The first step is the extraction of DNA from the organism known to have the trait of interest. The second step is gene cloning, which will isolate the gene of interest from the entire extracted DNA, followed by mass-production of the cloned gene in a host cell. Once it is cloned, the gene of interest is designed and packaged so that it can be controlled and properly expressed once inside the host plant. The modified gene will then be mass-produced in a host cell in order to make thousands of copies. When the gene package is ready, it can then be introduced into the cells of the plant being modified through a process called transformation. The most common methods used to introduce the gene package into plant cells include biolistic transformation (using a gene gun) or Agrobacterium-mediated transformation. Once the inserted gene is stable, inherited, and expressed in subsequent generations, then the plant is considered a transgenic. Backcross breeding is the final step in the genetic engineering process, where the transgenic crop is crossed with a variety that possesses important agronomic traits, and selected in order to obtain high quality plants that express the inserted gene in a desired manner.

The length of time in developing transgenic plant depends upon the gene, crop species, available resources, and regulatory approval. It may take 6-15 years before a new transgenic hybrid is ready for commercial release.

Commercially Available Crops Improved through Genetic Engineering

Transgenic crops have been planted in different countries for twenty years, starting from 1996 to 2015. About 179.7 million hectares was planted in 2015 to transgenic crops with high market value, such as herbicide tolerant soybean, maize, cotton, and canola; insect resistant maize, cotton, potato, and rice; and virus resistant squash and papaya. With genetic engineering, more than one trait can be incorporated or stacked into a plant. Transgenic crops with combined traits are also available commercially. These include herbicide tolerant and insect resistant maize, soybean and cotton.

New and Future Initiatives in Crop Genetic Engineering

To date, commercial GM crops have delivered benefits in crop production, but there are also a number of products in the pipeline which will make more direct contributions to food quality, environmental benefits, pharmaceutical production, and non-food crops. Examples of these products include: Rice with higher levels of iron and beta-carotene (an important micronutrient which is converted to vitamin A in the body); long life banana that ripens faster on the tree and can therefore be harvested earlier; tomatoes with high levels of flavonols, which are powerful antioxidants; arsenic-tolerant plants; edible vaccines from fruit and vegetables; and low lignin trees for paper making.

Genetic Resources in Crop Production

Genetic resources are the raw materials which scientists use to alter plant performance, and with which they hope to achieve the dramatic increase in crop production that will be required in the decades ahead.

Another factor of concern when considering biodiversity has been the emphasis on increasing yields by large scale production of superior clones. An obvious consequence of crop uniformity and neglect of wild related species, which often contain many useful traits, is a narrow gene pool and resultant vulnerability.

Papaya (Carica papaya) is an excellent example of a crop that is extremely vulnerable, but which is economically important in many tropical countries. In some of these countries, for example The Philippines, many farmers are already struggling with a subsistence existence and so the loss incurred by the failure of a cash crop causes real hardship.

The crop is threatened worldwide by a number of serious diseases, including viruses, namely Papaya Ring spot Virus (PRSV-P), fungal diseases including Phytophthora root, stem and fruit rots, and mycoplasmas. PRSV-P has devastated crop production in many regions and may cause many landraces or useful genotypes to be lost completely.

Extensive conventional crop breeding has achieved significantly improved fruit size and flavour and has been responsible for large improvements in crop yields but has been unsuccessful in addressing the current disease threats. However, the wild Carica relatives contain a huge pool of useful genes and valuable natural resistances to major papaya diseases have been observed.

Carica cauliflora, C. quercifolia, C. pubescens and C. stipulata have resistance to PRSV-P; Phytophthora-resistance has been found in C. goudotiana; and C. parviflora appears resistant to two mycoplasma diseases in field plantings at Redlands Research Station in Southeast Queensland, Australia.

Although the focus in collecting germplasm has been placed on plant species that are valuable as food sources, the case of papaya illustrates the value of other genetic resources. Documentation of wild relatives, landraces and their useful characteristics and in-situ or ex-situ collection requires much time, effort and funds.

However, the final benefits may be substantial, particularly as useful traits, such as resistance to a particular disease, potentially can be transferred to more than one crop species by gene transfer technology. The importance of germplasm conservation is being increasingly realised and many collections already exist.

The report on the State of World's Plant Genetic Resources records some of the difficulties being encountered, for example 'much of the plant genetic resources held in ex-situ collections are insufficiently and/ or poorly documented' and 'globally, governments and donor agencies have made insufficient provisions for on-going maintenance costs of conservation infrastructure' resulting in a 'steady deterioration of many facilities and their ability to perform even basic conservation functions', and, 'a high percentage of accessions that are held in these banks are in need of regeneration'.

These and other problems led to the adoption of The Global Plan of Action at the fourth International Technical Conference on Plant Genetic Resources for Food and Agriculture in June 1996.

Any attempt to systematically conserve genetic material on a worldwide basis will of course necessitate extensive government cooperation and funding. The task at hand is a daunting one but is complicated by the prevailing attitude of many governments in western countries to agricultural research.

Although they are signatories to all relevant conventions there is a growing reluctance to fund agricultural research. As funds dry up, competition causes available funding to go to those research areas promising significant results in the short term. Germplasm storage is not offering spectacular short term outcomes and is motivated by the need to take responsibility for the long term.

All countries need to value and be responsible for the conservation of genetic diversity. It is a natural resource, the value of which is still to be fully appreciated as science continues to unravel the mysteries of plant function at a cellular level and employ them to meet the constant stream of new and on-going pressures and difficulties facing crop production.

There is an urgent need to increase research into technologies that facilitate conservation of plant genetic resources. This is consistent with The Global Plan of Action.

Since the 1970's there has been a conscientious effort to collect germplasm. In 1995, Tao and Anishetty reported that worldwide more than 4 million accessions of crop plants and their relatives were stored in gene banks and that there were 256 gene banks with long-term storage facilities and 1227 institutions with ex-situ collections.

However, these statistics are rendered less impressive by two considerations. Firstly, 40% of these accessions are cereals (wheat represents 14%). No consistent effort has been applied to the collection of other major crops, while minor crops including fruit crops, have been largely neglected. Secondly, more than 90% of the accessions are stored as seed.

As a consequence, vegetative propagated species and species with recalcitrant seeds are greatly underrepresented. In addition, if we are to preserve diversity, collections must also include wild relatives of crop species.

As gene transfer techniques overcome the incompatibility between species, there is also a need to be mindful that non-crop species may also represent a valuable source of useful genes such as disease and insect resistance.

Although seed storage has been the conventional method of conserving germplasm, pollen storage and field collections have also been employed. Scientific advances in the field of cell and tissue culture systems have made available, as alternatives, conservation in vitro in either slow growth systems or cryopreservation.

More recently, developments in molecular biology offer the future prospect of DNA storage as libraries of either DNA sequence information or of DNA constructs. As reported in the State of the World's Plant Genetic Resources, one of the major concerns with germplasm storage is the maintenance of collections.

Stored germplasm requires careful characterisation as well as periodic regeneration to ensure viability. Techniques resulting from biotechnology offer improvements in this area. Molecular marker techniques enable characterisation at the DNA level, and an advantage of in vitro conservation is that viability is easily assessed visually.

As regular regeneration, usually annually, is essential, it is less likely to be overlooked. A major effort is still required in the evaluation and conservation of tropical and subtropical fruit species.

The development and maintenance of germplasm banks is of limited usefulness if the genetic material is not distributed and used worldwide to develop new and improved varieties of food plants. Hence The Global Plan of Action promoted the development of networks for the effective exchange and utilisation of germplasm.

It states that 'networks are important platforms for scientific exchange, information sharing, technology transfer, research collaboration, and for the determination and sharing of responsibilities for such activities as collecting, conservation, distribution, evaluation and genetic enhancement'.

FAO has supported and collaborated in the development of networks in the 1990's. Networks have been established on both a crop (mushroom, cactus pear, olive, citrus, nut tree) and regional basis.

The innovative aspect of the newly launched networks is the promotion of a coordinated approach to identifying, evaluating and conserving the genetic variability of selected crop species, with the aim of its utilisation for the improvement of cultivars and their adaptation to farmer's needs'.

There exists the potential to strengthen existing networks, develop new networks and to coordinate these regional and crop-based networks into an effective global system.

The chief aims of The Global Plan of Action are the conservation and sustainable use of genetic resources, by developing more effective conservation strategies at both the national and international level. A major advantage of crop-based networks is that they can facilitate linkages between conservation and sustainable utilisation of crop genetic resources.

A number of factors highlight the importance of a global approach. Plant Genetic Resources (PGR) is a common concern of humanity and all nations are both donors and users of PGR. The effective use of a wide range of PGR in all countries is an incentive to ensure their conservation and maintenance.

Thus the international community as a whole has both an interest in and a responsibility to conserve PGR and this is the basis for an effective integrated and rational Global Plan of Action. A related development which emphasises the need for a global approach is the proposed World Information and Early Warning System on Plant Genetic Resources (WIEWS).

The technical meeting on the methodology of WIEWS recommended the establishment of a global network, on a voluntary basis, under the auspices of FAO. WIEWS would provide a vehicle to register cases of genetic erosion both in in situ populations and ex situ collections.

However, it would require an accurate information base on the current status of PGR including crop species, local varieties and wild relatives. The Global Network would provide 'timely and precise information on PGR conservation and utilisation' contribute to the implementation and updating of The Global Plan of Action, provide an information flow between countries and develop an effective early warning system.

Genetic Engineering for Disease Resistance in Plants

Modern agriculture must provide sufficient nutrients to feed the world's growing population, which is projected to increase from 7.3 billion in 2015 to at least 9.8 billion by 2050. This goal is made even more challenging because of crop loss to diseases. Bacterial and fungal pathogens reduce crop yields by about 15% and viruses reduce yields by 3%. For some crops, such as potatoes, the loss caused by microbial infection is estimated to be as high as 30%. As an alternative to the application of chemical agents, researchers are altering the genetic composition of plants to enhance resistance to microbial infections.

Conventional breeding plays an essential role in crop improvement but usually entails growing and examining large populations of crops over multiple generations, a lengthy and labor-intensive process. Genetic engineering, which refers to the direct alteration of an organism's genetic material using biotechnology, possesses several advantages compared with conventional breeding. First, it enables the introduction, removal, modification, or fine-tuning of specific genes of interest with minimal undesired changes to the rest of the crop genome. As a result, crops exhibiting desired agronomic traits can be obtained in fewer generations compared with conventional breeding. Second, genetic engineering allows for interchange of genetic material across species. Thus, the raw genetic materials that can be exploited for this process is not restricted to the genes available within the species. Third, plant transformation during genetic engineering allows the introduction of new genes into vegetatively propagated crops such as banana (Musa sp.), cassava, and potato. These features make genetic engineering a powerful tool for enhancing resistance against plant pathogens.

Most cases of plant genetic engineering rely on conventional transgenic approaches or the more recent genome-editing technologies. In conventional transgenic methods, genes that encode desired agronomic traits are inserted into the genome at random locations through plant transformation. These methods typically result in varieties containing foreign DNA. In contrast, genome editing allows changes to the endogenous plant DNA, such as deletions, insertions, and replacements of DNA of various lengths at designated targets. Depending on the type of edits introduced, the product may or may not contain foreign DNA. In some areas of the world, including the United States, Argentina, and Brazil, genome-edited plants that do not contain foreign DNA are not subject to the additional regulatory measures applied to transgenic plants. Regardless of differences in regulatory policies, both conventional transgenic techniques and genome editing continue to be powerful tools for crop improvement.

Plants have evolved multilayered defense mechanisms against microbial pathogens. For example, preformed physical and physiological barriers and their reinforcements prevent potential pathogens from gaining access inside the cell. Plasma membrane-bound and intracellular immune receptors initiate defense responses upon the perception of pathogens either directly through physically interacting with pathogen-derived immunogens or indirectly by monitoring modifications of host targets incurred by pathogens. Plant-derived antimicrobial peptides and other compounds can suppress pathogenicity by direct detoxification or through inhibition of the activity of virulence factors. Plants also employ RNA interference to detect invading viral pathogens and target the viral RNA for cleavage.

On the other side, successful pathogens have evolved strategies to overcome the defense responses of their plant hosts. For instance, many bacterial and fungal pathogens produce and release cell-wall-degrading enzymes. Pathogens can also deliver effectors into the host cytoplasm; some of these effectors suppress host defenses and promote susceptibility. To counteract plant RNAi-based defenses, almost all plant viruses produce viral suppressors of RNAi. Some viruses also hijack the host RNAi system to silence host genes to promote viral pathogenicity.

Some aspects of host-microbe interactions provide opportunities for genetic engineering for disease resistance. For example, genes that encode proteins capable of breaking down mycotoxins or inhibiting the activity of cell-wall-degrading enzymes can be introduced into plants. Plants can be engineered to synthesize and secrete antimicrobial peptides or compounds that directly inhibit colonization. The RNAi machinery can be exploited to confer robust viral immunity by targeting viral RNA for degradation. Natural or engineered immune receptors that recognize diverse strains of a pathogen can be introduced individually or in combination for robust, broad-spectrum disease resistance. Essential defense hub regulatory genes can be reprogrammed to fine-tune defense responses. Susceptible host targets can be modified to evade recognition and manipulation by pathogens. Similarly, host decoy proteins, which serve to trap pathogens, can be modified through genetic engineering for altered specificity in pathogen recognition.

Increased knowledge regarding the molecular mechanism underlying plant-pathogen interactions and advancements in biotechnology have provided new opportunities for engineering resistance to microbial pathogens in plants. In this review, we describe recent breakthroughs in genetic engineering in plants for enhanced disease resistance and highlight several strategies that have proven successful in field trials.

Breakthroughs in Genetic Engineering for Disease Resistance in Plants

Researchers have long observed that transgenic plants expressing genes derived from viral pathogens often display immunity to the pathogen and its related strains. These results led to the hypothesis that ectopic expression of genes encoding wild-type or mutant viral proteins could interfere with the viral life cycle. More recent studies demonstrated that this immunity is mediated by RNAi, which plays a major role in antiviral defense in plants.

The detailed molecular mechanism of RNAi in antiviral defense has been described in several excellent reviews. Activation of RNAi has proven to be an effective approach for engineering resistance to viruses as they rely on the host cellular machinery to complete their life cycle. Most plant viruses contain single-stranded RNA as their genetic information. Double-stranded RNA (dsRNA) replicative intermediates often form during the replication of the viral genome mediated by RNA-dependent RNA polymerase, triggering RNAi in the host.

Transgenic overexpression of viral RNA often leads to the formation of dsRNA, which also triggers RNAi. This process is often known as cosuppression. By far, most existing cases of genetically engineered crops with resistance to viral pathogens that have been approved for cultivation and consumption were generated using this strategy. Notably, transgenic squash and papaya varieties created using this approach have been grown commercially in the United States for more than 20 years. Field data indicate that disease resistance achieved through RNAi-mediated strategies is highly durable. The RNAi strategy has also been effective in generating squash varieties that are resistant to multiple viral species These varieties were produced by gene stacking (i.e. transgenic expression of two distinct RNAi constructs targeted to different viruses.

The knowledge that dsRNA effectively induces RNAi inspired the design of transgenes encoding inverted repeat sequences, the transcripts of which form dsRNA. This strategy was used to develop a transgenic common bean variety exhibiting resistance to the DNA virus Bean golden mosaic virus. BGMV contains single-stranded DNA as its genetic material, which is converted to dsDNA in the host, transcribed, and translated to produce the essential proteins required for its replication. To generate small interfering RNA targeting the viral transcripts, sense and antisense sequences of part of the BGMV replication gene AC1 were directionally cloned into an intron-spliced hairpin RNA expression cassette. The resulting genetically engineered bean exhibited strong and robust resistance in greenhouse conditionsas well as field conditions. The BGMV-resistant bean was deregulated in 2011 in Brazil, which allows farmers to grow the crop. It is to date the only example of a deregulated genetically engineered crop showing resistance to a DNA virus. Compared with the cosuppression strategy, transgenic expression of an artificially designed inverted repeat allows simultaneous generation of heterogenous small interfering RNAs targeting multiple transcripts in a relatively simple manner.

The discovery of microRNAs (miRNAs), a class of endogenous noncoding regulatory RNAs led to further refinements of genetic engineering for viral resistance. The miRNA machinery has been exploited in engineering resistance to RNA viruses by replacing specific nucleotides in the miRNA-encoding genes to alter targeting specificity. Such artificial miRNAs have been used for engineering resistance to a wide range of plant viral pathogens including Turnip mosaic virus, Cucumber mosaic virus, Potato virus X, and Potato virus Y. These results suggest that artificial-miRNA-based antiviral strategies are highly promising. Disease resistance engineered by this strategy awaits future field tests.

Harnessing CRISPR-cas to Target Viral Pathogens Directly

In recent years, the bacterially derived clustered regularly interspaced short palindromic repeats (CRISPR)-CRISPR-associated (Cas) technology has proven to be a promising approach for engineering resistance against plant viruses. In many bacterial species, CRISPR-Cas acts as antiviral defense machinery. In this process, an RNA-guided nuclease (often a Cas protein) cleaves at specific target sites on the substrate viral DNA or

RNA, leading to their degradation. The specificity of the cleavage is governed by base complementarity between the CRISPR RNA and the target DNA or RNA molecules. A number of Cas proteins with sequence-specific nuclease activity have been identified. For example, the RNA-guided endonuclease Cas9 from Streptococcus pyogenes (SpCas9) induces double-stranded breaks in DNA in vivo and the RNA-guided RNases Cas13a from Leptotrichia shahii or Leptotrichia wadeitarget RNA in vivo. Cas9 from Francisella novicida (FnCas9) cleaves both DNA and RNA in vivo.

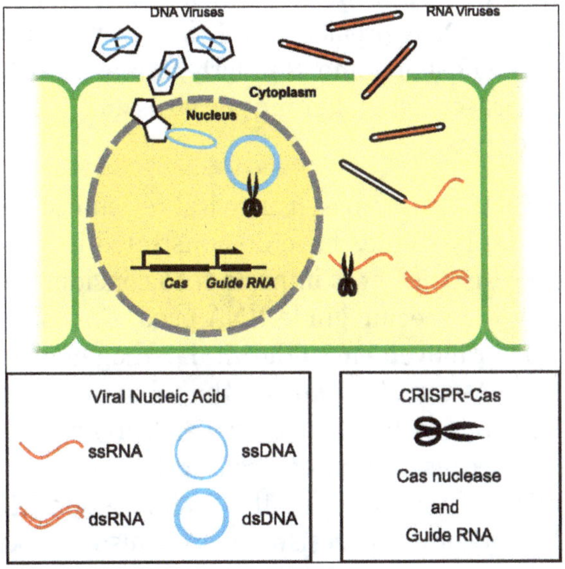

CRISPR-Cas platforms for engineering viral resistance in plants. Plant viruses complete their life cycles in the host cells. Viral particles with geminate capsids, and rod-shaped capsids are used as examples to illustrate DNA viruses and RNA viruses, respectively. One of the Cas nuclease genes (SpCas9, FnCas9, or LshCas13a) and the matching guide RNA(s) are expressed transgenically to generate sequence-specific ribonucleoproteins that target viral DNA or RNA substrates for degradation. SpCas9 targets nuclear dsDNA, while FnCas9 and LshCas13a have been used to target single-stranded RNA (ssRNA) substrates in the cytoplasm. For simplicity, only one generic Cas nuclease is drawn. ssDNA, single-stranded DNA.

The Tomato yellow leaf curl virus (TYLCV) belongs to Geminiviridae, a major family of plant viruses with single-stranded circular DNA genomes. The viral genome forms double-stranded intermediates during replication inside the host cell nuclei. Overexpression of SpCas9 and artificially designed guide RNAs targeting various regions of TYLCV conferred resistance to the virus in Nicotiana benthamiana and tomato. One potential caveat of this approach is that alterations to the viral DNA sequence may occur near the cleavage target due to DNA repair within the host cell. These mutations could shield the viral DNA from recognition by the guide RNA. Among all the guide RNAs used against TYLCV, the ones targeting the stem-loop sequence within the replication of origin were the most effective, possibly because of the reduced occurrence of viable escapee variants of the virus with mutations in this region. A separate lab study demonstrated that overexpression of SpCas9 and guide RNAs in N. benthamiana and Arabidopsis (Arabidopsis thaliana) conferred resistance to the geminivirus family member Beet severe curly top virus.

CRISPR-Cas has also been used to engineer resistance to RNA viruses, which comprise most known plant viral pathogens. For example, stable expression of the RNA-targeting nuclease Cas13a and the corresponding guide RNA in N. benthamiana conferred resistance to the RNA virus Turnip mosaic virus. Using Cas13a to target viral RNA substrates does not induce DNA breakage and thus would not introduce undesired off-target mutations to the host genome. Similarly, FnCas9 has been used to engineer resistance against RNA viruses Cucumber mosaic virus and Tobacco mosaic virus in N. benthamiana and Arabidopsis.

Although a field test of CRISPR-Cas-based antiviral resistance on crop species has not been reported, the laboratory studies have demonstrated its potential as an antiviral tool. To ensure durable resistance, it is important to consider potential viral evasions from the surveillance by the specific guide RNA used. Choosing genomic targets essential for the replication or movement of the viral pathogen minimizes viral evasions. Multiplexing the guide RNAs can also improve the robustness. In addition, it has been hypothesized that the use of Cas12a (also known as Cpf1) may reduce the occurrence of escapee viral variants because mutations caused by CRISPR-Cas12a are less likely to abloish the recognition of the target by the orginal guide RNA. Future efforts on improving CRISPR-Cas for antivirus resistance may also be focused on establishing an in planta adaptive immunity by exploiting the spacer acquisition machinery in the CRISPR-Cas adaptive immune system in prokaryotes.

Deploying Resistance Genes for Broad-spectrum and Durable Resistance

Pioneering genetic studies in plant-pathogen interaction using flax and flax rust fungus by Flor in the early 1940s established the classic "gene-for-gene" theory, which states that the outcome of any given plant-pathogen interaction is largely determined by a resistance (R) locus from the host and the matching avirulence factor (avr) from the pathogen. When both the R gene in the plant host and the cognate avr in the pathogen are present, the plant-pathogen interaction is incompatible and the host exhibits full resistance to the pathogen. The effectiveness of R gene-mediated resistance was first demonstrated by British scientist Rowland Biffen in wheat (Triticum sp.) breeding in the early twentieth century.

Since that time, numerous R genes have been cloned and introduced into varieties of the same species, across species boundaries and across genera. For example, the introduction of the maize R gene Rxo1 into rice conferred resistance to the bacterial streak pathogen Xanthomonas oryzae pv. oryzicola under lab conditions. In tomato, multi-year fields trials under commercial type growth conditions have demonstrated that tomatoes expressing the pepper Bs2 R gene confer robust resistance to Xanthomonas sp. causing the bacterial spot disease. Wheat transgenically expressing various alleles of the wheat resistance locus Pm3 exhibited race-specific resistance to powdery mildew in the field. Potatoes transgenically expressing wild potato R gene RB or Rpi-vnt1.1 display strong field resistance to Phytophthora infestans, the causal agent of potato late blight. Notably, transgenic potato expressing Rpi-vnt1.1 developed by J.R. Simplot is to date the only case of a genetically engineered crop with enhanced resistance to a nonviral pathogen that has been approved for commercial use.

Successful pathogens often evade detection by host R genes. Thus, disease resistance conferred by a single R gene often lacks durability in the field because pathogens can evolve to evade recognition by mutating the corresponding avr gene. For improved durability and to broaden the resistance spectrum, multiple R genes are often introduced simultaneously, which is commonly known as stacking. Resistance conferred by stacked R genes is predicted to be long lasting, as the evolution of a pathogen strain that could overcome resistance conferred by multiple R genes simultaneously is a low occurrence event. One approach to stack R genes is by cross breeding preexisting R loci. Breeders can then use marker-assisted selection to identify the progeny with the desired R gene composition. For example, bacterial blight in rice, caused by the bacterial pathogen Xanthomonas oryzae pv. oryzae (Xoo), is a serious disease in much of Asia and parts of Africa. Through cross breeding and marker-assisted selection, three R genes that confer resistance to bacterial blight in rice, Xa21, Xa5, and Xa13 were introduced into a deep-water rice cultivar called Jalmagna. The resulting line with the stacked R genes exhibited a higher level of field resistance to eight Xoo isolates tested. Although marker-assisted selection has largely improved the efficiency of the selection process, combining multiple loci through this approach can still be highly time consuming.

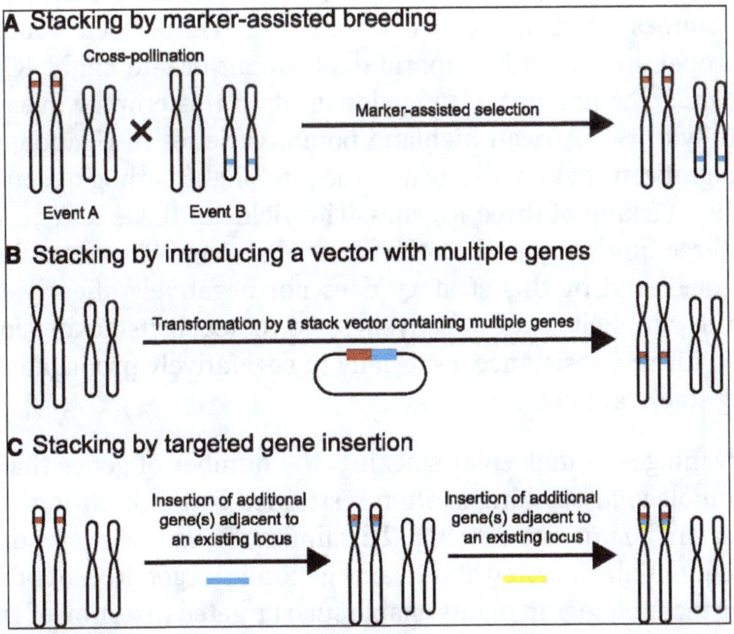

Methods of R gene stacking: A, Stacking by marker-assisted breeding is performed by cross pollinating individuals with existing trait loci followed by marker-assisted selection for progeny with combined trait loci. B, Stacking can be completed by combining multiple genes into a single stack vector and introducing them together as a single transgenic event. C, Stacking by targeted insertion aims at placing new gene(s) adjacent to an existing locus. This process can be performed iteratively to stack large numbers of genes. In B and C, the stacked genes are genetically linked and thus can be easily introduced into a different genetic background as a single locus through breeding.

As an alternative to gene stacking through marker-assisted selection, scientists can assemble multiple R gene cassettes onto one plasmid and then introduce this R gene cluster en bloc at a single genetic locus through plant transformation. This approach, called molecular stacking, simplifies the selection process, as all the R genes introduced this way are inherited as a single genetic locus. As an example of molecular stacking, Zhu et al. introduced molecular stacking of three broad-spectrum potato late blight R genes Rpi-sto1, Rpi-vnt1.1, and Rpi-blb3 through Agrobacterium-based transformation of a susceptible potato cultivar. The resulting triple gene transformants displayed broad-spectrum resistance equivalent to the sum of the strain-specific resistance conferred by all three individual Rpi genes under greenhouse conditions. In a related study, a single DNA fragment harboring Rpi-vnt1.1 and Rpi-sto1 was introduced into three different potato varieties through Agrobacterium-mediated transformation. The R gene-stacked lines showed broad-spectrum late blight resistance because of the introduction of both R genes. Notably, no foreign DNA such as a selectable marker gene was included in the inserted DNA fragment in the double gene study, which may reduce the number of regulatory approvals needed for the resulting products.

Resistance to the late blight pathogen in both the double gene-stacked and the triple gene-stacked potatoes mentioned above was confirmed under field conditions. It is estimated that proper spatial and temporal deployment of late blight R gene stacks in potatoes can reduce the amount of fungicide used on this crop by over 80%. Similarly, in a field study of two African highland potato varieties in Uganda, Ghislain et al. reported that significant field resistance to the late blight pathogen could be achieved by the molecular stacking of three R genes. The yields of these R gene stacked potato varieties were three times higher than the national average. These results demonstrate that resistance achieved by this strategy does not negatively affect yield. The above studies show the simplicity and effectiveness of molecular stacking for engineering broad-spectrum disease resistance, especially in vegetatively propagated crop species, where breeding stacks are not practical.

Despite the advantages of molecular stacking, the number of genes that can be introduced through molecular stacking is often restrained by the limit in the length of the DNA insert that can be put into a vector. This limitation can be overcome if DNA fragments can be sequentially inserted at the same genomic target. Recent breakthroughs in genome-editing technologies in plants enable such targeted insertion of DNA fragments in diverse crop species. This technology allows multiple R gene cassettes to be inserted at a single locus in multiple rounds. As genome-editing platforms in plants continue to be improved, the efficiency of targeted insertion, the size limit of the DNA inserts, and the number of applicable plant species are increasing. Future advancements in targeted gene insertion would offer new opportunities for stacking of large numbers of R genes and engineered viral resistance at a single locus for broad-spectrum, durable disease resistance and convenience in breeding.

Enriching the Known Repertoire of Immune Receptors

Plants possess immune receptors that perceive pathogens and trigger cellular defense responses. Discovery of novel immune receptors recognizing major virulence factors will enrich the repertoire of known immune receptor genes that may be deployed in the field. The nucleotide-binding leucine-rich repeats (NLR) family proteins comprise a major category of intracellular immune receptors conserved across the plant and animal kingdoms. A distinct hallmark for NLR-mediated defense is the onset of localized programmed cell death known as the hypersensitive response, which plays a crucial role in restricting the movement of pathogens. The hypersensitive response is often used as a marker to screen diverse plant germplasm for novel functional NLRs recognizing known effectors. During the screening, core virulence factors are delivered into plant leaves either as transiently expressed genes through Agro-infiltration or as genes expressed in laboratory bacterial strains with a functional secretion system.

Once a collection of germplasm exhibiting various degrees of resistance to a particular pathogen strain is identified, comparative genomics tools such as resistance gene enrichment sequencing can be applied to identify genomic variants in NLR genes that are linked to disease phenotypes. This leads to the cloning of new NLR genes and their potential deployment in crop protection through genetic engineering. For example, RenSeq was successfully applied in the accelerated identification and cloning of an anti-P. infestans NLR gene Rpi-amr3i from Solanum americanum, a wild potato relative. Transgenic expression of Rpi-amr3i in potato conferred full resistance to P. infestans in greenhouse conditions. In a more recent study, Chen et al. described the rapid mapping of a newly identified anti-potato-late-blight NLR gene Rpi-ver1 by RenSeq in the wild potato species Solanum verrucosum.

In addition to its application in potato research, RenSeq has also been employed in the cloning of wheat NLR resistance genes against the stem rust fungus Puccinia graminis f. sp. tritici. As an example, Arora et al. used RenSeq to survey accessions of a wild progenitor species of bread wheat, Aegilops tauschii ssp. strangulata for trait-linked. This led to the rapid cloning of four stem rust (Sr) resistance genes. In a related study, Steuernagel et al. applied RenSeq to a mutagenized hexaploid bread wheat population to identify mutations disrupting resistance to the stem rust fungus. The study revealed the identity of two stem rust NLR genes, Sr22 and Sr45, which confer resistance to commercially important races of the stem rust pathogen. Although future field experiments are required to evaluate the potential of the deployment of these newly identified NLR genes, the above lab studies demonstrate that RenSeq is a powerful tool to rapidly identify novel NLR genes.

In addition to the identification of useful immune receptors from diverse plant germplasm, researchers have attempted to engineer known immune receptors for new ligand specificity. For example, the fusion of the ectodomain of the Arabidopsis pattern recognition receptor EF-Tu receptor (EFR) to the intracellular domain of the phylogenetically related rice receptor Xa21 yielded a functional chimeric receptor in both rice and Arabidopsis. This receptor triggers defense markers when transgenic tissues are treated with elf18, the ligand of EFR, although whole-plant resistance to the microbe was weak or undetectable. Conversely, expression of a fusion receptor consisting of the ectodomain of Xa21 and the cytoplasmic domain of EFR in rice conferred robust resistance to Xoo expressing the cognate ligand of Xa21. Two related studies reported that chimeric receptors generated by fusing the rice chitin-binding protein Chitin Elicitor-Binding Protein and the intracellular protein kinase domain of the rice receptor-like kinase Xa21 or Pi-d2 confers disease resistance to the rice blast fungus. Although these studies have not been advanced to field trials, they demonstrate that domain swapping among immune receptors may be an attractive approach in engineering broadened recognition specificity.

NLRs often perceive pathogens indirectly by monitoring the modification of host target proteins by pathogen-derived virulence factors. Some of these host target proteins have evolved to serve as decoys that are targeted by virulence factors. For example, the NLR protein RESISTANCE TO PSEUDOMONAS SYRINGAE 5 (RPS5), activates defense responses in Arabidopsis. Defense is triggered when the plant decoy protein PBS1, a kinase, is cleaved by the protease AvrPphB secreted by the bacterial pathogen Pseudomonas syringae into the plant cell. Kim et al. altered the protease cleavage site in PBS1 to expand the resistance spectrum of RPS5 so that it could recognize other pathogens. For example, they found that they could remove the cleavage site of PBS1 that is recognized by the AvrPphB protease and replace it with cleavage site variants recognized by other pathogen proteases such as the AvrRpt2 protease from Pseudomonas syringae, or the Nla protease from Turnip mosaic virus. The engineered forms of PBS1 are cleaved in vivo by these proteases, which activates RPS5 defense in response to the corresponding pathogen strains. These successful examples make decoy modification an attractive approach to engineer resistance to new pathogens.

Altered specificity and activity of immune receptors can also be achieved by directed molecular evolution. During this process, immune receptor genes are subjected to mutagenesis to generate multiple variants, which are screened for their immune functions. The process can be performed iteratively for continuous improvement in the activity of the gene product. Two pioneering studies demonstrated the potential value of directed molecular evolution in expanding the pool of available immune receptor genes. In a search for gain-of-function mutants of the potato late blight NLR gene R3a, Segretin et al. screened a library of R3a variants generated through PCR-based mutagenesis and identified R3a mutants recognizing additional Avr3a isoforms from P. infestans and the related blight pathogen Phytophthora capsici. In a follow-up study, the newly identified mutation in R3a was transferred into its tomato ortholog I2 and expanded the recognition spectrum of the NLR encoded when transiently expressed in N. benthaminana.

Recently, a CRISPR-based mutagenesis platform known as EvolvR was developed for the directed evolution of a precisely defined window of genomic DNA in vivo. EvolvR employs the mutagenesis activity of a fusion of the Cas9 nickase and an error-prone DNA polymerase to introduce nucleotide diversification at specific genomic regions defined by guide RNAs. For example, EvolvR may potentially be used to develop novel alleles of the rice immune receptor Xa21 with the ability to recognize a broader array of Xoo strains, which secrete ligand variants. Variants of Xa21 receptors carrying amino acid substitutions predicted to alter ligand specificity can be screened for and introduced into susceptible rice cultivars. Future development of high-throughput functional screening methods to quickly identify gene variants conferring resistance phenotypes would broaden the application of directed molecular evolution for enhanced disease resistance in plants.

Modifying Susceptibility Genes to Attenuate Pathogenicity

Many plant pathogens hijack host genes to facilitate the infection process. These targeted host genes are often referred to as susceptibility genes. The modification or removal of host susceptibility genes may, therefore, be an effective strategy to achieve resistance by preventing their manipulation by the pathogens. RNAi and genome-editing platforms enable efficient modification of endogenous susceptibility genes in a relatively simple manner.

The highly conserved eukaryotic translation initiation factor 4E (eIF4E) in many plant species is manipulated by a number of plant viruses to facilitate the viral infection process. Naturally occurring recessive alleles of eIF4E confer viral resistance due to abolished protein-protein interaction with a specialized viral virulence protein. Based on this knowledge, Cavatorta et al. demonstrated that transgenic overexpression of a site-directed-mutagenized viral-resistant allele of the eIF4E conferred resistance to Potato virus Y in potato. In a more recent study, Chandrasekaran et al. knocked out the eIF4E gene in cucumber using CRISPR-Cas and obtained homozygous mutant plants that are resistant to a variety of viral pathogens under greenhouse conditions.

The SWEET family genes encode cross-membrane sugar transporters, many of which are exploited by plant pathogens for virulence. Often, when SWEET genes are activated, more sugar is transported outside of the cell and made available to the bacteria. For example, the promoters of rice SWEET11 and SWEET14 are targets of various transcriptional activator-like (TAL) effectors from Xoo. Li et al. used a TAL effector-like nuclease to mutate predicted TAL effector-binding sites within the SWEET14 promoter. The resulting rice exhibits enhanced resistance to at least two strains of Xoo due to the lack of induction of SWEET14 by the pathogens. In a follow-up study, Zhou et al. successfully generated rice lines carrying mutations in the promoter of SWEET11 using CRISPR-Cas. Mutations within the SWEET14 promoter in rice protoplasts by CRISPR-Cas have also been demonstrated. In addition to the genome-editing approach, RNAi has been employed to knock down SWEET11 in rice for enhanced resistance to an Xoo strain carrying the cognate TAL effector.

The disease susceptibility gene for citrus bacterial canker disease, Citrus sinensis Lateral Organ Boundary1 (CsLOB1) encodes a transcription factor that regulates plant growth and development. Deletion of the effector binding element within the promoter of the susceptibility gene CsLOB1 by CRISPR-Cas conferred resistance to the bacterial canker pathogen Xanthomonas citri spp. citri (Xcc) in orange. Similarly, disrupting the coding region of CsLOB1 by CRISPR-Cas in grapefruit (Citrus paradisi) resulted in enhanced resistance to Xcc. In both studies, the plants were raised in controlled environmental conditions. The effectiveness of resistance gained through mutating CsLOB1 in citrus plants awaits further evaluation by field experiments.

Susceptibility genes are often conserved among plant species. Therefore, knowledge of susceptibility genes in one plant species can facilitate the discovery of new susceptibility genes in another plant species. The Arabidopsis susceptibility gene DOWNY MILDEW RESISTANT6 (DMR6) encodes an oxygenase, and its expression is up-regulated during pathogen infection. Through phylogeny and gene expression analysis, Thomazella et al. identified a single DMR6 ortholog in tomato (SlDMR6-1). Knocking out SlDMR6-1 in tomato using CRISPR-Cas led to increased resistance to the bacterial pathogen P. syringae under greenhouse conditions. In another study, Sun et al. selected 11 Arabidopsis genes that had been previously shown to correlate with susceptibility to one or more of six common plant pathogens. In a search for gene products with similar amino acid sequences within the potato genome, they identified 11 candidate susceptibility genes in potato. Silencing of six of these candidate genes individually in potato using RNAi conferred enhanced or complete resistance to the late blight pathogen in controlled environmental conditions. These studies show that advancements in Arabidopsis biology and identification of an increasing number of susceptibility genes have enabled the discovery of additional susceptibility genes in a large variety of crop species.

Modifying susceptibility genes to thwart virulence strategies of pathogens holds great potential in crop protection. Because a given allele that confers susceptibility to one pathogen may confer resistance to another or have other essential biological functions, it is important to evaluate the potential side-effects of modifying susceptibility genes. It remains to be determined whether the observed resistance in the instances discussed in the laboratory studies is robust or durable under field conditions or if the plants perform well in the field.

HERBICIDE TOLERANT CROPS

Genetic engineering (GE) refers to techniques used to manipulate the genetic composition of an organism by adding specific genes. The enhancement of desired traits has traditionally been undertaken through conventional plant breeding. GE crops are often broken down into two categories, herbicide tolerant and Plant-incorporated protectants (PIPs).Crops are also engineered or "stacked" to express multiple traits like crops that are resistant to multiple herbicides or are resistant to herbicides and incorporates insecticides.

Herbicide tolerant crops are designed to tolerate specific broad-spectrum herbicides, which kill the surrounding weeds, but leave the cultivated crop intact. Currently, the only varieties Cultivated in the U.S. are engineered to be tolerant to glyphosate. However, the U.S. Department of Agriculture (USDA) is currently in the process of deregulating other new varieties of crops that are resistant to 2, 4-D and other herbicides.

Glyphosate Tolerant Crops

Monsanto first introduced glyphosate-resistant soybean in 1996 and later introduced glyphosate-resistant corn in 1998. These crops, commonly called "Roundup Ready", have become ubiquitous in American agriculture with 93% of soybeans, 82% of cotton, and 85% of corn planted engineered to be glyphosate resistant. This increase in glyphosate-resistant crops has led to an increase in herbicide use, herbicide-resistant weeds (also known as "superweeds"), and numerous other environmental and human health impacts.

Resistance

Pesticide resistance, the ability of an organism to withstand a poison, is a predictable consequence of repeated pesticide use. How quickly pesticide resistance develops depends on: The frequency of use, the mechanisms of resistance, the genetics of the resistance mechanism, the size of the gene pool and how quickly the organisms reproduce.

- The production of three GE crops—cotton, soybeans and corn—had increased. This finding contradicts industry claims that the technology would reduce pesticide applications. According to the study, the emergence and spread of glyphosate-resistant weeds is by far the most important factor driving up herbicide use on land planted with GE crops. Glyphosate-tolerant weeds were practically unknown before the introduction of Roundup-tolerant crops in 1996. But heavy reliance on the herbicide Roundup, whose active ingredient is glyphosate, has placed weed populations under progressively intense and unprecedented selection pressure, triggering a perfect storm for the emergence of glyphosate-resistant weeds. "Resistant weeds have become a major problem for many farmers reliant on GE crops, and are now driving up the volume of herbicide needed each year by about 25 percent," Dr. Benbrook said.
- Glyphosate resistance issues have led farmers to request emergency exemptions to use untested herbicides on now glyphosate-resistant weeds. For example, in November 2012, the U.S. Environmental Protection Agency (EPA) granted an emergency exemption for the use of fluridone, an aquatic herbicide that has never undergone scrutiny for its effects on endangered species, on GE cotton crops in order to control resistant weeds.

Increased Herbicide Use

A recent USDA report found that herbicide use on GE corn increased from around 1.5 pounds per planted acre in 2001 to more than 2.0 pounds per planted acre in 2010. Herbicide use on non-GMO corn has remained relatively level during that same time frame.

Environmental Impacts

Loss of Habitat

The increased use of glyphosate-resistant crops has led to declines in pollinator habitat. Historically, for butterflies in the U.S., their key source of food, milkweed, was found in several key states where the butterfly feeds and breeds: Iowa, Minnesota, Wisconsin, Illinois, Indiana, parts of Ohio and the eastern Dakotas. Now fields have been planted with more than 120 million acres of corn and soybeans genetically engineered to be tolerant to glyphosate, as well as other herbicides, allowing farmers to use glyphosate to kill milkweed in the field. According to researchers, the utilization of these GE crops has all but eliminated milkweeds from these fields, thus eliminating the butterfly's source of food.

Glyphosate in the Environment

A 2011 study, "Occurrence and fate of the herbicide glyphosate and its degradate aminomethylphosphonic acid in the atmosphere, conducted by the U.S. Geological Survey (USGS) monitored glyphosate content in air and water samples in the states of Iowa and Mississippi across two growing seasons. The results show that glyphosate is detected 60-100% of the time in both air and rain samples. Its concentration in rainfall is found to be at higher levels than for any other previously monitored pesticide.

A second 2011 study, Fate and transport of glyphosate and aminomethylphosphonic acid in surface waters of agricultural basins, conducted by USGS monitored water concentrations of glyphosate. The study found glyphosate persists in streams throughout the growing season of GE crops in Iowa and Mississippi, but is generally not observed during other times of the year. The degradation product of glyphosate, aminomethylphosphonic acid (AMPA), which has a longer environmental lifetime, is also frequently detected in streams and rain.

Genetic Drift/Contamination

Allowing GE crops to be grown close to organic and non-GE conventional produce increases the riskof genetic cross-contamination, as pollen from GE crops has the potential to drift onto non-GE crops and produce offspring. Because GE crops are prohibited under organic standards organic farmers may suffer significant financial losses if certified organic crops become polluted with genetically-engineered pollen.

- A recent Survey produced by Food & Water Watch, Organic Farmers Pay the Price for GMO Contamination, found that a third of U.S. organic farmers have experienced problems in their fields due to the nearby use of GE crops, and over half of those growers have had loads of grain rejected because of unwitting GE contamination.
- In May of 2013, USDA announced that unapproved GE wheat was found growing in an Oregon wheat field. After this discovery Japan cancelled its order to buy U.S. western white wheat. Monsanto has not conducted field trials in Oregon since 2001 when it reportedly withdrew from the state.
- In September of 2013, the U.S. Department of Agriculture (USDA) refused to take action or investigate after it was confirmed that GE alfalfa contaminated non-GE alfalfa in Washington State. USDA claimed the contamination is a "commercial issue" and should be addressed by the marketplace and not the government.

Human Health Risks

The increased use of glyphosate on glyphosate resistant crops could lead to increases in human health problems. Glyphosate-formulated herbicides have been linked to numerous health problems including cancer, particularly non-Hodgkin's lymphoma in three separate peer-reviewed studies, ADHD, rhinitis, and hormone disruption. Short term health effects include lung congestion and increased breathing rates. Chronic exposures at levels above Maximum Contaminant Levels (MCL) are likely to produce kidney damage and reproductive effects.

2, 4-D Tolerant Crops

Recently, USDA released for public input its Draft Environmental Impact Statement (DEIS), which calls for the deregulation of GE corn and soybeans engineered to be tolerant to the herbicide 2,4-D. Much like glyphosate, these new varieties of GE corn and soybeans are set to usher in dramatic increases in 2, 4-D. Dow AgroSciences produces these new GE crops under the brand name "Enlist" which will be stacked with glyphosate resistance.

Increased use of 2, 4-D could have dramatic impacts on human and environmental health. Scientists around the world have reported increased cancer risks in association with its use, especially for non-Hodgkins Lymphoma (NHL). It is also neurotoxic, genotoxic, and an endocrine disruptor.

2, 4-D has a high potential to leach from soils and can be a potential ground water contaminate. Environmental monitoring detected the herbicide in streams, groundwater and even drinking water. Studies document 2, 4-D's negative impacts on a wide range of animals. In birds, 2, 4-D exposure reduced hatching success and caused birth defects. Toxic to fish, 2, 4-D can bio-accumulate inside the fish. 2, 4-D also is toxic to honey bees and earthworms.

GENETICALLY MODIFIED PLANT

Genetically modified plants have been engineered for scientific research, to create new colours in plants, deliver vaccines, and to create enhanced crops. Many plant cells are pluripotent, meaning that a single cell from a mature plant can be harvested and then under the right conditions form a new plant. This ability can be taken advantage of by genetic engineers; by selecting for cells that have been successfully transformed in an adult plant a new plant can then be grown that contains the transgene in every cell through a process known as tissue culture.

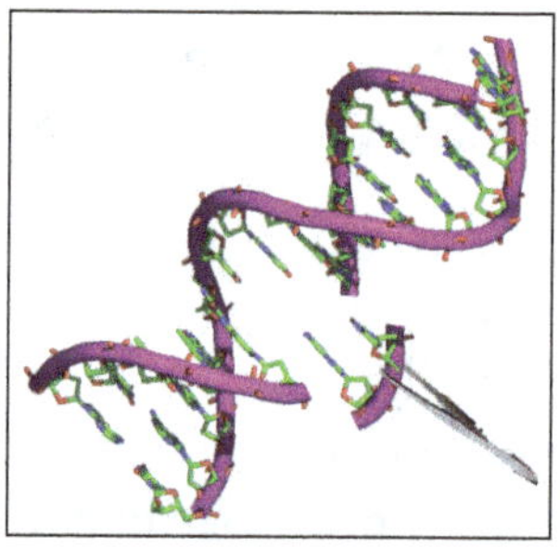

Much of the advances in the field genetic engineering has come from experimentation with tobacco. Major advances in tissue culture and plant cellular mechanisms for a wide range of plants has originated from systems developed in tobacco. It was the first plant to be genetically engineered and is considered a model organism for not only genetic engineering, but a range of other fields. As such the transgenic tools and procedures are well established making it one of the easiest plants to transform. Another major model organism relevant to genetic engineering is Arabidopsis thaliana. Its small genome and short life cycle makes it easy to manipulate and it contains many homolgs to important crop species. It was the first plant sequenced, has abundant bioinformatic resources and can be transformed by simply dipping a flower in a transformed Agrobacterium solution.

In research, plants are engineered to help discover the functions of certain genes. The simplest way to do this is to remove the gene and see what phenotype develops compared to the wild type form. Any differences are possibly the result of the missing gene. Unlike mutagenisis, genetic engineering allows targeted removal without disrupting other genes in the organism. Some genes are only expressed in certain tissue, so reporter genes, like GUS, can be attached to the gene of interest allowing visualisation of the location. Other ways to test a gene is to alter it slightly and then return it to the plant and see if it still has the same effect on phenotype. Other strategies include attaching the gene to a strong promoter and see what happens when it is over expressed, forcing a gene to be expressed in a different location or at different developmental stages.

Ornamental

Suntory "blue" rose.

Some genetically modified plants are purely ornamental. They are modified for lower color, fragrance, flower shape and plant architecture. The first genetically modified ornamentals commercialised altered colour. Carnations were released in 1997, with the most popular genetically modified organism, a blue rose (actually lavender or mauve) created in 2004. The roses are sold in Japan, the United States, and Canada. Other genetically modified ornamentals include Chrysanthemum and Petunia. As well as increasing aesthetic value there are plans to develop ornamentals that use less water or are resistant to the cold, which would allow them to be grown outside their natural environments.

Conservation

It has been proposed to genetically modify some plant species threatened by extinction to be resistant invasive plants and diseases, such as the emerald ash borer in North American and the fungal disease, Ceratocystis platani, in European plane trees. The papaya ringspot virus (PRSV) devastated papaya trees in Hawaii in the twentieth century until transgenic papaya plants were given pathogen-derived resistance. However, genetic modification for conservation in plants remains mainly speculative. A unique concern is that a transgenic species may no longer bear enough resemblance to the original species to truly claim that the original species is being conserved. Instead, the transgenic species may be genetically different enough to be considered a new species, thus diminishing the conservation worth of genetic modification.

Crops

Genetically modified crops are genetically modified plants that are used in agriculture. The first crops provided are used for animal or human food and provide resistance to certain pests, diseases, environmental conditions, spoilage or chemical treatments (e.g. resistance to a herbicide). The second generation of crops aimed to improve the quality, often by altering the nutrient profile. Third generation genetically modified crops can be used for non-food purposes, including the production of pharmaceutical agents, biofuels, and other industrially useful goods, as well as for bioremediation.

There are three main aims to agricultural advancement; increased production, improved conditions for agricultural workers and sustainability. GM crops contribute by improving harvests through reducing insect pressure, increasing nutrient value and tolerating different abiotic stresses. Despite this potential, as of 2018, the commercialised crops are limited mostly to cash crops like cotton, soybean, maize and canola and the vast majority of the introduced traits provide either herbicide tolerance or insect resistance. Soybeans accounted for half of all genetically modified crops planted in 2014. Adoption by farmers has been rapid, between 1996 and 2013, the total surface area of land cultivated with GM crops increased by a factor of 100, from 17,000 square kilometers (4,200,000 acres) to 1,750,000 km^2 (432 million acres). Geographically though the spread has been very uneven, with strong growth in the Americas and parts of Asia and little in Europe and Africa. Its socioeconomic spread has been more even, with approximately 54% of worldwide GM crops grown in developing countries in 2013.

Food

The majority of GM crops have been modified to be resistant to selected herbicides, usually a glyphosate or glufosinate based one. Genetically modified crops engineered to resist herbicides are now more available than conventionally bred resistant varieties; in the USA 93% of soybeans and most of the GM maize grown is glyphosate tolerant. Most currently available genes used to engineer insect resistance come from the Bacillus thuringiensis bacterium. Most are in the form of delta endotoxin genes known as cry proteins, while a few use the genes that encode for vegetative insecticidal proteins. The only gene commercially used to provide insect protection that does not originate from B. thuringiensis is the Cowpea trypsin inhibitor (CpTI). CpTI was first approved for use cotton in 1999 and is currently undergoing trials in rice. Less than one percent of GM crops contained other traits, which include providing virus resistance, delaying senescence, modifying flower colour and altering the plants composition. Golden rice is the most well-known GM crop that is aimed at increasing nutrient value. It has been engineered with three genes that biosynthesise beta-carotene, a precursor of vitamin A, in the edible parts of rice. It is intended to produce a fortified food to be grown and consumed in areas with a shortage of dietary vitamin A. a deficiency which each year is estimated to kill 670,000 children under the age of 5 and cause an additional 500,000 cases of irreversible childhood blindness. The original golden rice produced 1.6μg/g of the carotenoids, with further development increasing this 23 times. In 2018 it gained its first approvals for use as food.

Biopharmaceuticals

Plants and plant cells have been genetically engineered for production of biopharmaceuticals in bioreactors, a process known as Pharming. Work has been done with duckweed Lemna minor, the algae Chlamydomonas reinhardtii and the moss Physcomitrella patens. Biopharmaceuticals produced include cytokines, hormones, antibodies, enzymes and vaccines, most of which are accumulated in the plant seeds. Many drugs also contain natural plant ingredients and the pathways that lead to their production have been genetically altered or transferred to other plant species to produce greater volume and better products. Other options for bioreactors are biopolymers and biofuels. Unlike bacteria, plants can modify the proteins post-translationally, allowing them to make more complex molecules.They also pose less risk of being contaminated. Therapeutics have been cultured in transgenic carrot and tobacco cells, including a drug treatment for Gaucher's disease.

Vaccines

Vaccine production and storage has great potential in transgenic plants. Vaccines are expensive to produce, transport and administer, so having a system that could produce them locally would allow greater access to poorer and developing areas. As well as purifying vaccines expressed in plants it is also possible to produce edible vaccines in plants. Edible vaccines stimulate the immune system when ingested to protect against certain diseases. Being stored in plants reduces the long-term cost as they can be disseminated without the need for cold storage, don't need to be purified and have long term stability. Also being housed within plant cells provides some protection from the gut acids upon digestion. However the cost of developing, regulating and containing transgenic plants is high, leading to most current plant-based vaccine development being applied to veterinary medicine, where the controls are not as strict.

GENETICALLY MODIFIED CROPS

Genetically modified crops (GM crops) are plants used in agriculture, the DNA of which has been modified using genetic engineering methods. In most cases, the aim is to introduce a new trait to the plant which does not occur naturally in the species. Examples in food crops include resistance to certain pests, diseases, environmental conditions, reduction of spoilage, resistance to chemical treatments (e.g. resistance to a herbicide), or improving the nutrient profile of the crop. Examples in non-food crops include production of pharmaceutical agents, biofuels, and other industrially useful goods, as well as for bioremediation.

Farmers have widely adopted GM technology. Acreage increased from 1.7 million hectares in 1996 to 185.1 million hectares in 2016, some 12% of global cropland. As of 2016, major crop (soybean, maize, canola and cotton) traits consist of herbicide tolerance (95.9 million hectares) insect resistance (25.2 million hectares), or both (58.5 million hectares). In 2015, 53.6 million ha of GM maize were under cultivation (almost 1/3 of the maize crop). GM maize outperformed its predecessors: yield was 5.6 to 24.5% higher with less mycotoxins (−28.8%), fumonisin (−30.6%) and thricotecens (−36.5%). Non-target organisms were unaffected, except for Braconidae, represented by a parasitoid of European corn borer, the target of Lepidoptera active Bt maize. Biogeochemical parameters such as lignin content did not vary, while biomass decomposition was higher.

A 2014 meta-analysis concluded that GM technology adoption had reduced chemical pesticide use by 37%, increased crop yields by 22%, and increased farmer profits by 68%. This reduction in pesticide use has been ecologically beneficial, but benefits may be reduced by overuse. Yield gains and pesticide reductions are larger for insect-resistant crops than for herbicide-tolerant crops. Yield and profit gains are higher in developing countries than in developed countries.

There is a scientific consensus that currently available food derived from GM crops poses no greater risk to human health than conventional food, but that each GM food needs to be tested on a case-by-case basis before introduction. Nonetheless, members of the public are much less likely than scientists to perceive GM foods as safe. The legal and regulatory status of GM foods varies by country, with some nations banning or restricting them, and others permitting them with widely differing degrees of regulation.

However, opponents have objected to GM crops on grounds including environmental impacts, food safety, whether GM crops are needed to address food needs, whether they are sufficiently accessible to farmers in developing countries and concerns over subjecting crops to intellectual property law. Safety concerns led 38 countries, including 19 in Europe, to officially prohibit their cultivation.

Methods

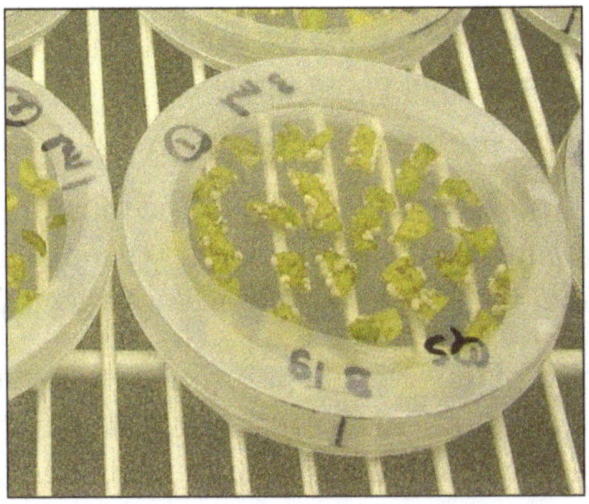

Plants (Solanum chacoense) being transformed using agrobacterium.

Genetically engineered crops have genes added or removed using genetic engineering techniques, originally including gene guns, electroporation, microinjection and agrobacterium. More recently, CRISPR and TALEN offered much more precise and convenient editing techniques.

Gene guns (also known as biolistics) "shoot" (direct high energy particles or radiations against) target genes into plant cells. It is the most common method. DNA is bound to tiny particles of gold or tungsten which are subsequently shot into plant tissue or single plant cells under high pressure. The accelerated particles penetrate both the cell wall and membranes. The DNA separates from the metal and is integrated into plant DNA inside the nucleus. This method has been applied successfully for many cultivated crops, especially monocots like wheat or maize, for which transformation using Agrobacterium tumefaciens has been less successful. The major disadvantage of this procedure is that serious damage can be done to the cellular tissue.

Agrobacterium tumefaciens-mediated transformation is another common technique. Agrobacteria are natural plant parasites. Their natural ability to transfer genes provides another engineering method. To create a suitable environment for themselves, these Agrobacteria insert their genes into plant hosts, resulting in a proliferation of modified plant cells near the soil level (crown gall). The genetic information for tumor growth is encoded on a mobile, circular DNA fragment (plasmid). When Agrobacterium infects a plant, it transfers this T-DNA to a random site in the plant genome. When used in genetic engineering the bacterial T-DNA is removed from the bacterial plasmid and replaced with the desired foreign gene. The bacterium is a vector, enabling transportation of foreign genes into plants. This method works especially well for dicotyledonous plants like potatoes, tomatoes, and tobacco. Agrobacteria infection is less successful in crops like wheat and maize.

Electroporation is used when the plant tissue does not contain cell walls. In this technique, "DNA enters the plant cells through miniature pores which are temporarily caused by electric pulses." Microinjection is used to directly inject foreign DNA into cells.

Plant scientists, backed by results of modern comprehensive profiling of crop composition, point out that crops modified using GM techniques are less likely to have unintended changes than are conventionally bred crops.

In research tobacco and Arabidopsis thaliana are the most frequently modified plants, due to well-developed transformation methods, easy propagation and well-studied genomes. They serve as model organisms for other plant species.

Introducing new genes into plants requires a promoter specific to the area where the gene is to be expressed. For instance, to express a gene only in rice grains and not in leaves, an endosperm-specific promoter is used. The codons of the gene must be optimized for the organism due to codon usage bias.

Types of Modifications

Transgenic

Transgenic maize containing a gene from the bacteria Bacillus thuringiensis.

Transgenic plants have genes inserted into them that are derived from another species. The inserted genes can come from species within the same kingdom (plant to plant), or between kingdoms (for example, bacteria to plant). In many cases the inserted DNA has to be modified slightly in order to be correctly and efficiently expressed in the host organism. Transgenic plants are used to express proteins, like the cry toxins from B. thuringiensis, herbicide-resistant genes, antibodies, and antigens for vaccinations. A study led by the European Food Safety Authority (EFSA) also found viral genes in transgenic plants.

Transgenic carrots have been used to produce the drug Taliglucerase alfa which is used to treat Gaucher's disease. In the laboratory, transgenic plants have been modified to increase photosynthesis (currently about 2% at most plants versus the theoretic potential of 9–10%). This is possible by changing the rubisco enzyme (i.e. changing C_3 plants into C_4 plants), by placing the rubisco in a carboxysome, by adding CO_2 pumps in the cell wall, or by changing the leaf form or size. Plants have been engineered to exhibit bioluminescence that may become a sustainable alternative to electric lighting.

Cisgenic

Cisgenic plants are made using genes found within the same species or a closely related one, where conventional plant breeding can occur. Some breeders and scientists argue that cisgenic modification is useful for plants that are difficult to crossbreed by conventional means (such as potatoes), and that plants in the cisgenic category should not require the same regulatory scrutiny as transgenics.

Subgenic

Genetically modified plants can also be developed using gene knockdown or gene knockout to alter the genetic makeup of a plant without incorporating genes from other plants. In 2014, Chinese researcher Gao Caixia filed patents on the creation of a strain of wheat that is resistant to powdery mildew. The strain lacks genes that encode proteins that repress defenses against the mildew. The researchers deleted all three copies of the genes from wheat's hexaploid genome. Gao used the TALENs and CRISPR gene editing tools without adding or changing any other genes. No field trials were immediately planned. The CRISPR technique has also been used by Penn State researcher Yinong Yang to modify white button mushrooms (Agaricus bisporus) to be non-browning, and by DuPont Pioneer to make a new variety of corn.

Multiple Trait Integration

With multiple trait integration, several new traits may be integrated into a new crop.

Economics

GM food's economic value to farmers is one of its major benefits, including in developing nations. A 2010 study found that Bt corn provided economic benefits of $6.9 billion over the previous 14 years in five Midwestern states. The majority ($4.3 billion) accrued to farmers producing non-Bt corn. This was attributed to European corn borer populations reduced by exposure to Bt corn, leaving fewer to attack conventional corn nearby. Agriculture economists calculated that "world surplus [increased by] $240.3 million for 1996. Of this total, the largest share (59%) went to U.S. farmers. Seed company Monsanto received the next largest share (21%), followed by US consumers (9%), the rest of the world (6%), and the germplasm supplier, Delta & Pine Land Company of Mississippi (5%)."

According to the International Service for the Acquisition of Agri-biotech Applications (ISAAA), in 2014 approximately 18 million farmers grew biotech crops in 28 countries about 94% of the farmers were resource-poor in developing countries. 53% of the global biotech crop area of 181.5 million hectares was grown in 20 developing countries. PG Economics comprehensive 2012 study concluded that GM crops increased farm incomes worldwide by $14 billion in 2010, with over half this total going to farmers in developing countries.

Critics challenged the claimed benefits to farmers over the prevalence of biased observers and by the absence of randomized controlled trials. The main Bt crop grown by small farmers in developing countries is cotton. A 2006 review of Bt cotton findings by agricultural economists concluded, "the overall balance sheet, though promising, is mixed. Economic returns are highly variable over years, farm type, and geographical location".

In 2013 the European Academies Science Advisory Council (EASAC) asked the EU to allow the development of agricultural GM technologies to enable more sustainable agriculture, by employing fewer land, water, and nutrient resources. EASAC also criticizes the EU's "time-consuming and expensive regulatory framework" and said that the EU had fallen behind in the adoption of GM technologies.

Participants in agriculture business markets include seed companies, agrochemical companies, distributors, farmers, grain elevators and universities that develop new crops/traits and whose agricultural extensions advise farmers on best practices. According to a 2012 review based on data from the late 1990s and early 2000s, much of the GM crop grown each year is used for livestock feed and increased demand for meat leads to increased demand for GM feed crops. Feed grain usage as a percentage of total crop production is 70% for corn and more than 90% of oil seed meals such as soybeans. About 65 million metric tons of GM corn grains and about 70 million metric tons of soybean meals derived from GM soybean become feed.

In 2014 the global value of biotech seed was US$15.7 billion; US$11.3 billion (72%) was in industrial countries and US$4.4 billion (28%) was in the developing countries. In 2009, Monsanto had $7.3 billion in sales of seeds and from licensing its technology; DuPont, through its Pioneer subsidiary, was the next biggest company in that market. As of 2009, the overall Roundup line of products including the GM seeds represented about 50% of Monsanto's business.

Some patents on GM traits have expired, allowing the legal development of generic strains that include these traits. For example, generic glyphosate-tolerant GM soybean is now available. Another impact is that traits developed by one vendor can be added to another vendor's proprietary strains, potentially increasing product choice and competition. The patent on the first type of Roundup Ready crop that Monsanto produced (soybeans) expired in 2014 and the first harvest of off-patent soybeans occurs in the spring of 2015. Monsanto has broadly licensed the patent to other seed companies that include the glyphosate resistance trait in their seed products. About 150 companies have licensed the technology, including Syngenta and DuPont Pioneer.

Yield

In 2014, the largest review yet concluded that GM crops' effects on farming were positive. The meta-analysis considered all published English-language examinations of the agronomic and economic impacts between 1995 and March 2014 for three major GM crops: Soybean, maize, and cotton. The study found that herbicide-tolerant crops have lower production costs, while for insect-resistant crops the reduced pesticide use was offset by higher seed prices, leaving overall production costs about the same.

Yields increased 9% for herbicide tolerance and 25% for insect resistant varieties. Farmers who adopted GM crops made 69% higher profits than those who did not. The review found that GM crops help farmers in developing countries, increasing yields by 14 percentage points.

The researchers considered some studies that were not peer-reviewed and a few that did not report sample sizes. They attempted to correct for publication bias, by considering sources beyond academic journals. The large data set allowed the study to control for potentially confounding variables such as fertilizer use. Separately, they concluded that the funding source did not influence study results.

Traits

GM crops grown today, or under development, have been modified with various traits. These traits include improved shelf life, disease resistance, stress resistance, herbicide resistance, pest resistance, production of useful goods such as biofuel or drugs, and ability to absorb toxins and for use in bioremediation of pollution.

Recently, research and development has been targeted to enhancement of crops that are locally important in developing countries, such as insect-resistant cowpea for Africa and insect-resistant brinjal (eggplant).

Genetically modified King Edward potato (right) next to King Edward which has not been genetically modified (left).

Extended Shelf Life

The first genetically modified crop approved for sale in the U.S. was the FlavrSavr tomato, which had a longer shelf life. First sold in 1994, FlavrSavr tomato production ceased in 1997. It is no longer on the market.

In November 2014, the USDA approved a GM potato that prevents bruising.

In February 2015 Arctic Apples were approved by the USDA, becoming the first genetically modified apple approved for US sale. Gene silencing was used to reduce the expression of polyphenol oxidase (PPO), thus preventing enzymatic browning of the fruit after it has been sliced open. The trait was added to Granny Smith and Golden Delicious varieties. The trait includes a bacterial antibiotic resistance gene that provides resistance to the antibiotic kanamycin. The genetic engineering involved cultivation in the presence of kanamycin, which allowed only resistant cultivars to survive. Humans consuming apples do not acquire kanamycin resistance, per arcticapple.com. The FDA approved the apples in March 2015.

Improved Photosynthesis

Plants use non-photochemical quenching to protect them from excessive amounts of sunlight. Plants can switch on the quenching mechanism almost instantaneously, but it takes much longer for it to switch off again. During the time that it is switched off, the amount of energy that is wasted increases. A genetic modification in three genes allows to correct this (in a trial with tobacco plants). As a result, yields were 14-20% higher, in terms of the weight of the dry leaves harvested. The plants had larger leaves, were taller and had more vigorous roots.

Another improvement that can be made on the photosynthesis process (with C3 pathway plants) is on photorespiration. By inserting the C4 pathway into C3 plants, productivity may increase by as much as 50% for cereal crops, such as rice.

Improved Nutritional Value

Edible Oils

Some GM soybeans offer improved oil profiles for processing. Camelina sativa has been modified to produce plants that accumulate high levels of oils similar to fish oils.

Vitamin Enrichment

Golden rice, developed by the International Rice Research Institute (IRRI), provides greater amounts of vitamin A targeted at reducing vitamin A deficiency. As of January 2016, golden rice has not yet been grown commercially in any country.

Toxin Reduction

A genetically modified cassava under development offers lower cyanogen glucosides and enhanced protein and other nutrients (called BioCassava).

In November 2014, the USDA approved a potato that prevents bruising and produces less acrylamide when fried. They do not employ genes from non-potato species. The trait was added to the Russet Burbank, Ranger Russet and Atlantic varieties.

Stress Resistance

Plants have been engineered to tolerate non-biological stressors, such as drought, frost, and high soil salinity. In 2011, Monsanto's DroughtGard maize became the first drought-resistant GM crop to receive US marketing approval.

Drought resistance occurs by modifying the plant's genes responsible for the mechanism known as the crassulacean acid metabolism (CAM), which allows the plants to survive despite low water levels. This holds promise for water-heavy crops such as rice, wheat, soybeans and poplar to accelerate their adaptation to water-limited environments. Several salinity tolerance mechanisms have been identified in salt-tolerant crops. For example, rice, canola and tomato crops have been genetically modified to increase their tolerance to salt stress.

Herbicides

Glyphosate

As of 1999, the most prevalent GM trait was glyphosate-tolerance. Glyphosate (the active ingredient in Roundup and other herbicide products) kills plants by interfering with the shikimate pathway in plants, which is essential for the synthesis of the aromatic amino acids phenylalanine, tyrosine, and tryptophan. The shikimate pathway is not present in animals, which instead obtain aromatic amino acids from their diet. More specifically, glyphosate inhibits the enzyme 5-enolpyruvylshikimate-3-phosphate synthase (EPSPS).

This trait was developed because the herbicides used on grain and grass crops at the time were highly toxic and not effective against narrow-leaved weeds. Thus, developing crops that could withstand spraying with glyphosate would both reduce environmental and health risks, and give an agricultural edge to the farmer.

Some micro-organisms have a version of EPSPS that is resistant to glyphosate inhibition. One of these was isolated from an Agrobacterium strain CP4 (CP4 EPSPS) that was resistant to glyphosate. The CP4 EPSPS gene was engineered for plant expression by fusing the 5' end of the gene to a chloroplast transit peptide derived from the petunia EPSPS. This transit peptide was used because it had shown previously an ability to deliver bacterial EPSPS to the chloroplasts of other plants. This CP4 EPSPS gene was cloned and transfected into soybeans.

The plasmid used to move the gene into soybeans was PV-GMGTO4. It contained three bacterial genes, two CP4 EPSPS genes, and a gene encoding beta-glucuronidase (GUS) from Escherichia coli as a marker. The DNA was injected into the soybeans using the particle acceleration method. Soybean cultivar A5403 was used for the transformation.

Bromoxynil

Tobacco plants have been engineered to be resistant to the herbicide bromoxynil.

Glufosinate

Crops have been commercialized that are resistant to the herbicide glufosinate, as well. Crops engineered for resistance to multiple herbicides to allow farmers to use a mixed group of two, three, or four different chemicals are under development to combat growing herbicide resistance.

2, 4-D

In October 2014 the US EPA registered Dow's Enlist Duo maize, which is genetically modified to be resistant to both glyphosate and 2,4-D, in six states. Inserting a bacterial aryloxyalkanoate dioxygenase gene, aad1 makes the corn resistant to 2,4-D. The USDA had approved maize and soybeans with the mutation in September 2014.

Dicamba

Monsanto has requested approval for a stacked strain that is tolerant of both glyphosate and dicamba. The request includes plans for avoiding herbicide drift to other crops. Significant damage to other non-resistant crops occurred from dicamba formulations intended to reduce volatilization drifting when sprayed on resistant soybeans in 2017. The newer dicamba formulation labels specify to not spray when average wind speeds are above 10–15 miles per hour (16–24 km/h) to avoid particle drift, average wind speeds below 3 miles per hour (4.8 km/h) to avoid temperature inversions, and rain or high temperatures are in the next day forecast. However, these conditions typically only occur during June and July for a few hours at a time.

Pest Resistance

Insects

Tobacco, corn, rice and some other crops have been engineered to express genes encoding for insecticidal proteins from Bacillus thuringiensis (Bt). The introduction of Bt crops during the period between 1996 and 2005 has been estimated to have reduced the total volume of insecticide active ingredient use in the United States by over 100 thousand tons. This represents a 19.4% reduction in insecticide use.

In the late 1990s, a genetically modified potato that was resistant to the Colorado potato beetle was withdrawn because major buyers rejected it, fearing consumer opposition.

Viruses

Papaya, potatoes, and squash have been engineered to resist viral pathogens such as cucumber mosaic virus which, despite its name, infects a wide variety of plants. Virus resistant papaya were developed in response to a papaya ringspot virus (PRV) outbreak in Hawaii in the late 1990s. They incorporate PRV DNA. By 2010, 80% of Hawaiian papaya plants were genetically modified.

Potatoes were engineered for resistance to potato leaf roll virus and Potato virus Y in 1998. Poor sales led to their market withdrawal after three years.

Yellow squash that were resistant to at first two, then three viruses were developed, beginning in the 1990s. The viruses are watermelon, cucumber and zucchini/courgette yellow mosaic. Squash was the second GM crop to be approved by US regulators. The trait was later added to zucchini.

Many strains of corn have been developed in recent years to combat the spread of Maize dwarf mosaic virus, a costly virus that causes stunted growth which is carried in Johnson grass and spread by aphid insect vectors. These strands are commercially available although the resistance is not standard among GM corn variants.

By-products

Drugs

In 2012, the FDA approved the first plant-produced pharmaceutical, a treatment for Gaucher's Disease. Tobacco plants have been modified to produce therapeutic antibodies.

Biofuel

Algae is under development for use in biofuels. Researchers in Singapore were working on GM jatropha for biofuel production. Syngenta has USDA approval to market a maize trademarked Enogen that has been genetically modified to convert its starch to sugar for ethanol. Some trees have been genetically modified to either have less lignin, or to express lignin with chemically labile bonds. Lignin is the critical limiting factor when using wood to make bio-ethanol because lignin limits the accessibility of cellulose microfibrils to depolymerization by enzymes. Besides with trees, the chemically labile lignin bonds are also very useful for cereal crops such as maize, barley, and oats.

Materials

Companies and labs are working on plants that can be used to make bioplastics. Potatoes that produce industrially useful starches have been developed as well. Oilseed can be modified to produce fatty acids for detergents, substitute fuels and petrochemicals.

Bioremediation

Scientists at the University of York developed a weed (Arabidopsis thaliana) that contains genes from bacteria that could clean TNT and RDX-explosive soil contaminants in 2011. 16 million hectares in the US (1.5% of the total surface) are estimated to be contaminated with TNT and RDX. However A. thaliana was not tough enough for use on military test grounds. Modifications in 2016 included switchgrass and bentgrass.

Genetically modified plants have been used for bioremediation of contaminated soils. Mercury, selenium and organic pollutants such as polychlorinated biphenyls (PCBs).

Marine environments are especially vulnerable since pollution such as oil spills are not containable. In addition to anthropogenic pollution, millions of tons of petroleum annually enter the marine environment from natural seepages. Despite its toxicity, a considerable fraction of petroleum oil entering marine systems is eliminated by the hydrocarbon-degrading activities of microbial communities. Particularly successful is a recently discovered group of specialists, the so-called hydrocarbonoclastic bacteria (HCCB) that may offer useful genes.

Asexual reproduction

Crops such as maize reproduce sexually each year. This randomizes which genes get propagated to the next generation, meaning that desirable traits can be lost. To maintain a high-quality crop, some farmers purchase seeds every year. Typically, the seed company maintains two inbred varieties and crosses them into a hybrid strain that is then sold. Related plants like sorghum and gamma grass are able to perform apomixis, a form of asexual reproduction that keeps the plant's DNA intact. This trait is apparently controlled by a single dominant gene, but traditional breeding has been unsuccessful in creating asexually-reproducing maize. Genetic engineering offers another route to this goal. Successful modification would allow farmers to replant harvested seeds that retain desirable traits, rather than relying on purchased seed.

Genetic modifications to some crops also exist, which make it easier to process the crop, i.e. by growing in a more compact form. Also, some crops (such as tomatoes) have been genetic modified to contain no seed at all.

Development

The number of USDA-approved field releases for testing grew from 4 in 1985 to 1,194 in 2002 and averaged around 800 per year thereafter. The number of sites per release and the number of gene constructs (ways that the gene of interest is packaged together with other elements) – have rapidly increased since 2005. Releases with agronomic properties (such as drought resistance) jumped from 1,043 in 2005 to 5,190 in 2013. As of September 2013, about 7,800 releases had been approved for corn, more than 2,200 for soybeans, more than 1,100 for cotton, and about 900 for potatoes. Releases were approved for herbicide tolerance (6,772 releases), insect resistance (4,809), product quality such as flavor or nutrition (4,896), agronomic properties like drought resistance (5,190), and virus/fungal resistance (2,616). The institutions with the most authorized field releases include Monsanto with 6,782, Pioneer/DuPont with 1,405, Syngenta with 565, and USDA's Agricultural Research Service with 370. As of September 2013 USDA had received proposals for releasing GM rice, squash, plum, rose, tobacco, flax, and chicory.

Farming Practices

Bacillus Thuringiensis

Constant exposure to a toxin creates evolutionary pressure for pests resistant to that toxin. Over-reliance on glyphosate and a reduction in the diversity of weed management practices allowed the spread of glyphosate resistance in 14 weed species in the US, and in soybeans.

To reduce resistance to Bacillus thuringiensis (Bt) crops, the 1996 commercialization of transgenic cotton and maize came with a management strategy to prevent insects from becoming resistant. Insect resistance management plans are mandatory for Bt crops. The aim is to encourage a large population of pests so that any (recessive) resistance genes are diluted within the population. Resistance lowers evolutionary fitness in the absence of the stressor, Bt. In refuges, non-resistant strains outcompete resistant ones.

With sufficiently high levels of transgene expression, nearly all of the heterozygotes (S/s), i.e., the largest segment of the pest population carrying a resistance allele, will be killed before maturation, thus preventing transmission of the resistance gene to their progeny. Refuges (i. e., fields of nontransgenic plants) adjacent to transgenic fields increases the likelihood that homozygous resistant (s/s) individuals and any surviving heterozygotes will mate with susceptible (S/S) individuals from the refuge, instead of with other individuals carrying the resistance allele. As a result, the resistance gene frequency in the population remains lower.

Complicating factors can affect the success of the high-dose/refuge strategy. For example, if the temperature is not ideal, thermal stress can lower Bt toxin production and leave the plant more susceptible. More importantly, reduced late-season expression has been documented, possibly resulting from DNA methylation of the promoter. The success of the high-dose/refuge strategy has successfully maintained the value of Bt crops. This success has depended on factors independent of management strategy, including low initial resistance allele frequencies, fitness costs associated with resistance, and the abundance of non-Bt host plants outside the refuges.

Companies that produce Bt seed are introducing strains with multiple Bt proteins. Monsanto did this with Bt cotton in India, where the product was rapidly adopted. Monsanto has also; in an attempt to simplify the process of implementing refuges in fields to comply with Insect Resistance Management(IRM) policies and prevent irresponsible planting practices; begun marketing seed bags with a set proportion of refuge (non-transgenic) seeds mixed in with the Bt seeds being sold. Coined "Refuge-In-a-Bag" (RIB), this practice is intended to increase farmer compliance with refuge requirements and reduce additional labor needed at planting from having separate Bt and refuge seed bags on hand. This strategy is likely to reduce the likelihood of Bt-resistance occurring for corn rootworm, but may increase the risk of resistance for lepidopteran corn pests, such as European corn borer. Increased concerns for resistance with seed mixtures include partially resistant larvae on a Bt plant being able to move to a susceptible plant to survive or cross pollination of refuge pollen on to Bt plants that can lower the amount of Bt expressed in kernels for ear feeding insects.

Herbicide Resistance

Best management practices (BMPs) to control weeds may help delay resistance. BMPs include applying multiple herbicides with different modes of action, rotating crops, planting weed-free seed, scouting fields routinely, cleaning equipment to reduce the transmission of weeds to other fields, and maintaining field borders. The most widely planted GM crops are designed to tolerate herbicides. By 2006 some weed populations had evolved to tolerate some of the same herbicides. Palmer amaranth is a weed that competes with cotton. A native of the southwestern US, it traveled east and was first found resistant to glyphosate in 2006, less than 10 years after GM cotton was introduced.

Plant Protection

Farmers generally use less insecticide when they plant Bt-resistant crops. Insecticide use on corn farms declined from 0.21 pound per planted acre in 1995 to 0.02 pound in 2010. This is consistent with the decline in European corn borer populations as a direct result of Bt corn and cotton. The establishment of minimum refuge requirements helped delay the evolution of Bt resistance. However, resistance appears to be developing to some Bt traits in some areas.

Tillage

By leaving at least 30% of crop residue on the soil surface from harvest through planting, conservation tillage reduces soil erosion from wind and water, increases water retention, and reduces soil degradation as well as water and chemical runoff. In addition, conservation tillage reduces the carbon footprint of agriculture. A 2014 review covering 12 states from 1996 to 2006, found that a 1% increase in herbicde-tolerant (HT) soybean adoption leads to a 0.21% increase in conservation tillage and a 0.3% decrease in quality-adjusted herbicide use.

Production

In 2013, GM crops were planted in 27 countries; 19 were developing countries and

8 were developed countries. 2013 was the second year in which developing countries grew a majority (54%) of the total GM harvest. 18 million farmers grew GM crops; around 90% were small-holding farmers in developing countries.

Country	2013– GM planted area (million hectares)	Biotech crops
US	70.1	Maize, Soybean, Cotton, Canola, Sugarbeet, Alfalfa, Papaya, Squash
Brazil	40.3	Soybean, Maize, Cotton
Argentina	24.4	Soybean, Maize, Cotton
India	11.0	Cotton
Canada	10.8	Canola, Maize, Soybean, Sugarbeet
Total	175.2	----

The United States Department of Agriculture (USDA) reports every year on the total area of GM crop varieties planted in the United States. According to National Agricultural Statistics Service, the states published in these tables represent 81–86 percent of all corn planted area, 88–90 percent of all soybean planted area, and 81–93 percent of all upland cotton planted area (depending on the year).

Global estimates are produced by the International Service for the Acquisition of Agri-biotech Applications (ISAAA) and can be found in their annual reports, "Global Status of Commercialized Transgenic Crops".

Farmers have widely adopted GM technology. Between 1996 and 2013, the total surface area of land cultivated with GM crops increased by a factor of 100, from 17,000 square kilometers (4,200,000 acres) to 1,750,000 km² (432 million acres). 10% of the world's arable land was planted with GM crops in 2010. As of 2011, 11 different transgenic crops were grown commercially on 395 million acres (160 million hectares) in 29 countries such as the US, Brazil, Argentina, India, Canada, China, Paraguay, Pakistan, South Africa, Uruguay, Bolivia, Australia, Philippines, Myanmar, Burkina Faso, Mexico and Spain. One of the key reasons for this widespread adoption is the perceived economic benefit the technology brings to farmers. For example, the system of planting glyphosate-resistant seed and then applying glyphosate once plants emerged provided farmers with the opportunity to dramatically increase the yield from a given plot of land, since this allowed them to plant rows closer together. Without it, farmers had to plant rows far enough apart to control post-emergent weeds with mechanical tillage. Likewise, using Bt seeds means that farmers do not have to purchase insecticides, and then invest time, fuel, and equipment in applying them. However critics have disputed whether yields are higher and whether chemical use is less, with GM crops.

In the US, by 2014, 94% of the planted area of soybeans, 96% of cotton and 93% of corn were genetically modified varieties. Genetically modified soybeans carried herbicide-tolerant traits only, but maize and cotton carried both herbicide tolerance and insect protection traits (the latter largely Bt protein). These constitute "input-traits" that are aimed to financially benefit the producers, but may have indirect environmental benefits and cost benefits to consumers. The Grocery Manufacturers of America estimated in 2003 that 70–75% of all processed foods in the U.S. contained a GM ingredient.

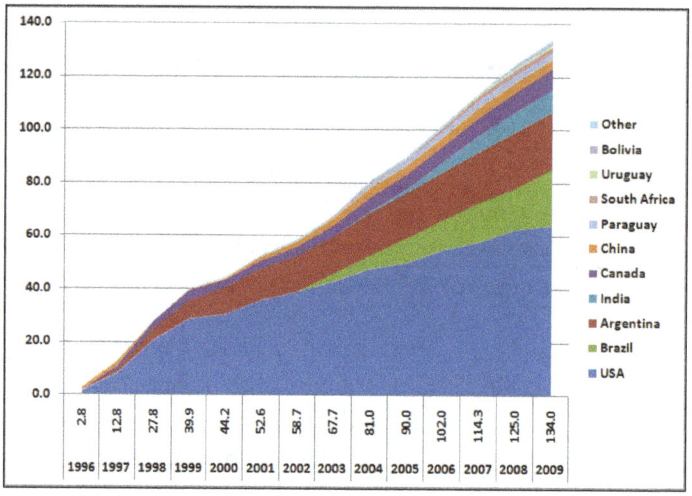

Land area used for genetically modified crops by country, in millions of hectares. In 2011, the land area used was 160 million hectares, or 1.6 million square kilometers.

Europe grows relatively few genetically engineered crops with the exception of Spain, where one fifth of maize is genetically engineered, and smaller amounts in five other countries. The EU had a 'de facto' ban on the approval of new GM crops, from 1999 until 2004. GM crops are now regulated by the EU. In 2015, genetically engineered crops are banned in 38 countries worldwide, 19 of them in Europe. Developing countries grew 54 percent of genetically engineered crops in 2013.

In recent years GM crops expanded rapidly in developing countries. In 2013 approximately 18 million farmers grew 54% of worldwide GM crops in developing countries. 2013's largest increase was in Brazil (403,000 km^2 versus 368,000 km^2 in 2012).

According to the 2013 ISAAA brief: "A total of 36 countries (35 + EU-28) have granted regulatory approvals for biotech crops for food and/or feed use and for environmental release or planting since 1994. A total of 2,833 regulatory approvals involving 27 GM crops and 336 GM events (NB: an "event" is a specific genetic modification in a specific species) have been issued by authorities, of which 1,321 are for food use (direct use or processing), 918 for feed use (direct use or processing) and 599 for environmental release or planting. Japan has the largest number (198), followed by the U.S.A. (165, not including "stacked" events), Canada (146), Mexico (131), South Korea (103), Australia (93), New Zealand (83), European Union (71 including approvals that have expired or under renewal process), Philippines (68), Taiwan (65), Colombia (59), China (55) and South Africa (52). Maize has the largest number (130 events in 27 countries), followed by cotton (49 events in 22 countries), potato (31 events in 10 countries), canola (30 events in 12 countries) and soybean (27 events in 26 countries).

Direct genetic engineering has been controversial since its introduction. Most, but not all of the controversies are over GM foods rather than crops per se. GM foods are the subject of protests, vandalism, referenda, legislation, court action and scientific disputes. The controversies involve consumers, biotechnology companies, governmental regulators, non-governmental organizations and scientists.

Opponents have objected to GM crops on multiple grounds including environmental impacts, food safety, whether GM crops are needed to address food needs, whether they are sufficiently accessible to farmers in developing countries and concerns over subjecting crops to intellectual property law. Secondary issues include labeling, the behavior of government regulators, the effects of pesticide use and pesticide tolerance.

A significant environmental concern about using genetically modified crops is possible cross-breeding with related crops, giving them advantages over naturally occurring varieties. One example is a glyphosate-resistant rice crop that crossbreeds with a weedy relative, giving the weed a competitive advantage. The transgenic hybrid had higher rates of photosynthesis, more shoots and flowers, and more seeds than the non-transgenic hybrids. This demonstrates the possibility of ecosystem damage by GM crop usage.

There is a scientific consensus that currently available food derived from GM crops poses no greater risk to human health than conventional food, but that each GM food needs to be tested on a case-by-case basis before introduction. Nonetheless, members of the public are much less likely than scientists to perceive GM foods as safe. The legal and regulatory status of GM foods varies by country, with some nations banning or restricting them, and others permitting them with widely differing degrees of regulation.

No reports of ill effects from GM food have been documented in the human population. GM crop labeling is required in many countries, although the United States Food and Drug Administration does not, nor does it distinguish between approved GM and non-GM foods. The United States enacted a law that requires labeling regulations to be issued by July 2018. It allows indirect disclosure such as with a phone number, bar code, or web site.

Advocacy groups such as Center for Food Safety, Union of Concerned Scientists, Greenpeace and the World Wildlife Fund claim that risks related to GM food have not been adequately examined and managed, that GM crops are not sufficiently tested and should be labelled, and that regulatory authorities and scientific bodies are too closely tied to industry. Some studies have claimed that genetically modified crops can cause harm; a 2016 review that reanalyzed the data from six of these studies found that their statistical methodologies were flawed and did not demonstrate harm, and said that conclusions about GM crop safety should be drawn from "the totality of the evidence instead of far-fetched evidence from single studies".

Genetically Modified Canola

Genetically modified canola is a genetically modified crop. The first strain, Roundup Ready canola, was developed by Monsanto for tolerance to glyphosate, the active ingredient in the commonly used herbicide Roundup.

Genetic Modification

Glyphosate is a broad-spectrum herbicide, which is used to kill weeds and grasses which are known to compete with commercial crops grown around the world. The first product came onto the market in the 1970's under the name 'Roundup'. Plants which are exposed to glyphosate are unable to produce aromatic amino acids and in turn die.

To produce the Roundup Ready canola, two genes were introduced into the canola genome. One is a gene derived from the common soil bacterium Agrobacterium strain CP4, that encodes for the EPSPS enzyme. The other is a gene from the Ochrobactrum anthropi strain LBAA, which encodes for the enzyme glyphosate oxidase (GOX). The CP4 EPSPS enzyme imparts high tolerance to glyphosate, so the plants can still create aromatic amino acids even after glyphosate is applied. GOX helps break down glyphosate within the plant.

Controversy exists over the use of food and other goods derived from genetically modified crops instead of from conventional crops, and other uses of genetic engineering in food production. The dispute involves consumers, biotechnology companies, governmental regulators, nongovernmental organizations, and scientists. The key areas of controversy related to GMO foods are whether they should be labeled, the role of government regulators, the objectivity of scientific research and publication, the effect of GM crops on health and the environment, the effect on pesticide resistance, the impact of GM crops for farmers, and the role of GM crops in feeding the world population.

There is a scientific consensus that currently available food derived from GM crops poses no greater risk to human health than conventional food, but that each GM food needs to be tested on a case-by-case basis before introduction. Nonetheless, members of the public are much less likely than scientists to perceive GM foods as safe. The legal and regulatory status of GM foods varies by country, with some nations banning or restricting them, and others permitting them with widely differing degrees of regulation.

Advocacy groups such as Greenpeace, the Non-GMO Project, and Organic Consumers Association say that risks of GM food have not been adequately identified and managed, and have questioned the objectivity of regulatory authorities. They have expressed concerns about the objectivity of regulators and rigor of the regulatory process, about contamination of the non-GM food supply, about effects of GMOs on the environment and nature, and about the consolidation of control of the food supply in companies that make and sell GMOs.

Resistances Problems

Due to the heavy reliance of glyphosate in agriculture, resistance to this chemical is a problem and is prevalent throughout Australia, the USA, and Canada.

Roundup canola has also emerged as a weed in other crops due to its glyphosate resistance. This is due to canola seed being able to be dormant in the soil for up to 10 years. In California, it has become a significant problem in this way because of the restrictions on phenoxy herbicides being used in the state due to crops such as the sensitivity of cotton and grapes to this chemical.

Genetically Modified Rice

Genetically modified rice are rice strains that have been genetically modified (also called genetic engineering). Rice plants have been modified to increase micronutrients such as vitamin A, accelerate photosynthesis, tolerate herbicides, resist pests, increase grain size, generate nutrients, flavours or produce human proteins.

The natural movement of genes across species, often called horizontal gene transfer or lateral gene transfer, can also occur with rice through gene transfer mediated by natural vectors. Transgenic events between rice and Setaria millet have been identified. The cultivation and use of genetically modified varieties of rice remains controversial and is not approved in some countries.

Rice plants being used for genetic modification.

In 2000, the first two GM rice varieties both with herbicide-resistance, called LLRice60 and LLRice62, were approved in the United States. Later, these and other types of herbicide-resistant GM rice were approved in Canada, Australia, Mexico and Colombia.

However, none of these approvals triggered commercialization. Reuters reported in 2009 that China had granted biosafety approval to GM rice with pest resistance, but that strain was not commercialized. As of December 2012 GM rice was not widely available for production or consumption. Advocates claim that since rice is a staple crop across the world, improvements have potential to alleviate hunger, malnutrition and poverty.

Traits

Herbicide Resistance

In 2000-2001 Monsanto researched adding glyphosate tolerance to rice but did not attempt to bring a variety to market. Bayer's line of herbicide resistant rice is known as LibertyLink. LibertyLink rice is resistant to glufosinate (the active chemical in Liberty herbicide). Bayer CropScience is attempting to get their latest variety (LL62) approved for use in the EU. The strain is approved for use in the US but is not in large-scale use. Clearfield rice was bred by selection from variations created in environments known to cause accelerated rates of mutations. This variety tolerates imidazole herbicides. It was bred by traditional breeding techniques that are not considered to be genetic engineering. Clearfield is also crossbred with higher yielding varieties to produce an overall hardier plant.

Nutritional Value

Golden rice with higher concentrations of Vitamin A was originally created by Ingo Potrykus and his team. This genetically modified rice is capable of producing beta-carotene in the endosperm (grain) which is a precursor for vitamin A. Syngenta was involved in the early development of Golden Rice and held some intellectual property that it donated to non-profit groups including the International Rice Research Institute (IRRI) to develop on a non-profit basis. The scientific details of the rice were first published in Science Magazine in 2000.

Golden Rice grains (right) compared to regular rice grains (left).

Golden Rice plants being grown in greenhouse.

The World Health Organization stated that iron deficiency affects 30% of the world's population. Research scientists from the Australian Centre for Plant Functional Genomics (ACPFG) and IRRI to are working to increase the amount of iron in rice. They have modified three populations of rice by over expressing the genes OsNAS1, OsNAS2 or OsNAS3. The research team found that nicotianamine, iron, and zinc concentration levels increased in all three populations relative to controls.

Ventria Bioscience uses a proprietary system known as Express Tec for producing recombinant human proteins in rice grains. Their most notable variety produces human Lactoferrin and Lysozyme. These two proteins are produced naturally in human breast milk and are used globally in infant formula and rehydration products.

Pest Resistance

BT rice is modified to express the cryIA(b) gene of the Bacillus thuringiensis bacterium. The gene confers resistance to a variety of pests including the rice borer through the production of endotoxins. The Chinese Government is doing field trials on insect resistant cultivars. The benefit of BT rice is that farmers do not need to spray their crops with pesticides to control fungal, viral, or bacterial pathogens. Conventional rice is sprayed three to four times per growing season to control pests. Other benefits include increased yield and revenue from crop cultivation. China approved the rice for large-scale use as of 2009.

Allergy Resistance

Researchers in Japan are attempting to develop hypoallergenic rice cultivars. Researchers are trying to repress the formation of allergen AS-Albumin.

Japanese researchers tested genetically modified rice on macaque monkeys that would prevent allergies to cedar pollen, which causes hay fever. Cedar allergy symptoms include itchy eyes, sneezing and other serious allergic reactions. The modified rice contains seven proteins within cedar pollen to block these symptoms. Takaiwa is conducting human clinical trials with this riceHuman blood protein.

Human serum albumin (HSA) is a blood protein in human blood plasma. It is used to treat severe burns, liver cirrhosis and hemorrhagic shock. It is also used in donated blood and is in short supply around the world. In China, scientists modified brown rice as a cost-effective way to produce HSA protein. The Chinese scientists put recombinant HSA protein promoters into 25 rice plants using Agrobacterium. Out of the 25 plants, nine contained the HSA protein. The genetically modified brown rice had the same amino acid sequence as HSA. They called this protein Oryza sativa recombinant HSA. The modified rice was transparent.

C4 Photosynthesis

In 2015 a consortium of 12 laboratories in eight countries developed a cultivar that displayed a rudimentary form of C4 photosynthesis (C4P) to boost growth by capturing carbon dioxide and concentrated it in specialized leaf cells. C4P is the reason corn and sugarcane grow so rapidly. Engineering C4 photosynthesis into rice could increase yields per hectare by roughly 50 percent. The current cultivar still relies primarily on C3 photosynthesis. To get them to completely adopt C4P, the plants must produce specialized cells in a precise arrangement: One set of cells to capture the carbon dioxide and to surround other cells that concentrate it. Some (possibly dozens of) genes involved in producing these cells remain to be identified. Other C3P crops that could exploit such knowledge include wheat, potatoes, tomatoes, apples and soybeans.

Controversies surround genetically modified organisms on several levels, including ethics, environmental impact, food safety, product labeling, role in meeting world food requirements, intellectual property, and role in industrial agriculture.

Legal Issues

In the summer of 2006, the USDA detected trace amounts of LibertyLink variety 601 in rice shipments ready for export. LL601 was not approved for food purposes. Bayer applied for deregulation of LL601 in late July and the USDA granted deregulation status in November 2006. The contamination led to a dramatic dip in rice futures markets with losses to farmers who grew rice for export. Approximately 30 percent of rice production and 11,000 farmers in Arkansas, Louisiana, Mississippi, Missouri and Texas were affected. In June 2011 Bayer agreed to pay 750 million dollars in damages and lost harvests. Japan and Russia suspended rice imports from the U.S., while Mexico and the European Union imposed strict testing. The contamination occurred between 1998 and 2001. The exact cause of the contamination was not discovered.

The Chinese government does not issue commercial usage licenses for genetically modified rice. All GM rice is approved for research only. Pu, et al., stated that rice engineered to produce human blood protein (HSA) requires a lot of modified rice to be grown. This raised environmental safety concerns about gene flow. They argued that this would not be a problem because rice is a self-pollinating crop, and their test showed less than 1% of the modified gene transferred in pollination. Another study suggested that insect-mediated gene flow may be higher than previously assumed.

Genetically Modified Soybean

A genetically modified soybean is a soybean (Glycine max) that has had DNA introduced into it using genetic engineering techniques. In 1996 the first genetically modified soybean was introduced to the U.S. market, by Monsanto. In 2014, 90.7 million hectares of GM soy were planted worldwide, 82% of the total soy cultivation area.

Examples of Transgenic Soybeans

The genetic makeup of a soybean gives it a wide variety of uses, thus keeping it in high demand. First, manufacturers only wanted to use transgenics to be able to grow more soy at a minimal cost to meet this demand, and to fix any problems in the growing process, but they eventually found they could modify the soybean to contain healthier components, or even focus on one aspect of the soybean to produce in larger quantities. These phases became known as the first and second generation of genetically modified (GM) foods. As Peter Celec describes, "benefits of the first generation of GM foods were oriented towards the production process and companies, the second generation of GM foods offers, on contrary, various advantages and added value for the consumer", including "improved nutritional composition or even therapeutic effects."

Ready Soybean

Roundup Ready soybeans (The first variety was also known as GTS 40-3-2 (OECD UI: MON-04032-6)) are a series of genetically engineered varieties of glyphosate-resistant soybeans produced by Monsanto.

Glyphosate kills plants by interfering with the synthesis of the essential amino acids phenylalanine, tyrosine and tryptophan. These amino acids are called "essential" because animals cannot make them; only plants and micro-organisms can make them and animals obtain them by eating plants.

Plants and microorganisms make these amino acids with an enzyme that only plants and lower organisms have, called 5-enolpyruvylshikimate-3-phosphate synthase (EPSPS). EPSPS is not present in animals, which instead obtain aromatic amino acids from their diet.

Roundup Ready Soybeans express a version of EPSPS from the CP4 strain of the bacteria Agrobacterium tumefaciens, expression of which is regulated by an enhanced 35S promoter (E35S) from cauliflower mosaic virus (CaMV), a chloroplast transit peptide (CTP4) coding sequence from Petunia hybrida, and a nopaline synthase (nos 3') transcriptional termination element from Agrobacterium tumefaciens. The plasmid with EPSPS and the other genetic elements mentioned above was inserted into soybean germplasm with a gene gun by scientists at Monsanto and Asgrow. The patent on the first generation of Roundup Ready soybeans expired in March 2015.

First approved commercially in the United States during 1994, GTS 40-3-2 was subsequently introduced to Canada in 1995, Japan and Argentina in 1996, Uruguay in 1997, Mexico and Brazil in 1998, and South Africa in 2001. GMO Soybean is also approved by the United Nations in 1999.

Detection

GTS 40-3-2 can be detected using both nucleic acid and protein analysis methods.

Generic GMO Soybeans

Following expiration of Monsanto's patent on the first variety of glyphosate-resistant Roundup Ready soybeans, development began on glyphosate-resistant generic soybeans. The first variety, developed at the University of Arkansas Division of Agriculture, came on the market in 2015. With a slightly lower yield than newer Monsanto varieties, it costs about 1/2 as much, and seeds can be saved for subsequent years. According to its creator it is adapted to conditions in Arkansas. Several other varieties are being bred by crossing the original variety of Roundup Ready soybeans with other soybean varieties.

Stacked Traits

Monsanto developed a glyphosate-resistant soybean that also expresses Cry1Ac protein from Bacillus thuringiensis and the glyphosate-resistance gene, which completed the Brazilian regulatory process in 2010.

Genetic Modification to Improve Soybean Oil

Soy has been genetically modified to improve the quality of soy oil. Soy oil has a fatty acid profile that makes it susceptible to oxidation, which makes it rancid, and this has limited its usefulness to the food industry. Genetic modifications increased the amount of oleic acid and stearic acid and decreased the amount of linolenic acid. By silencing, or knocking out, the delta 9 and delta 12 desaturases. DuPont Pioneer created a high oleic fatty acid soybean with levels of oleic acid greater than 80%, and started marketing it in 2010.

The regulation of genetic engineering concerns the approaches taken by governments to assess and manage the risks associated with the development and release of genetically modified crops. There are differences in the regulation of GM crops between countries, with some of the most marked differences occurring between the US and Europe. Soy beans are allowed a Maximum Residue Limit of glyphosate of 20 mg/Kg for international trade. Regulation varies in a given country depending on the intended use of the products of the genetic engineering. For example, a crop not intended for food use is generally not reviewed by authorities responsible for food safety.

There is a scientific consensus that currently available food derived from GM crops poses no greater risk to human health than conventional food, but that each GM food needs to be tested on a case-by-case basis before introduction. Nonetheless, members of the public are much less likely than scientists to perceive GM foods as safe. The legal and regulatory status of GM foods varies by country, with some nations banning or restricting them, and others permitting them with widely differing degrees of regulation.

A 2010 study found that in the United States, GM crops also provide a number of ecological benefits.

Critics have objected to GM crops on several grounds, including ecological concerns, and economic concerns raised by the fact these organisms are subject to intellectual property law. GM crops also are involved in controversies over GM food with respect to whether food produced from GM crops is safe and whether GM crops are needed to address the world's food needs. See the genetically modified food controversies article for discussion of issues about GM crops and GM food. These controversies have led to litigation, international trade disputes, and protests, and to restrictive legislation in most countries.

GENETICALLY MODIFIED FOOD

Genetically modified foods (GM foods), also known as genetically engineered foods (GE foods), or bioengineered foods are foods produced from organisms that have had changes introduced into their DNA using the methods of genetic engineering. Genetic engineering techniques allow for the introduction of new traits as well as greater control over traits when compared to previous methods, such as selective breeding and mutation breeding.

Commercial sale of genetically modified foods began in 1994, when Calgene first marketed its unsuccessful Flavr Savr delayed-ripening tomato. Most food modifications have primarily focused on cash crops in high demand by farmers such as soybean, corn, canola, and cotton. Genetically modified crops have been engineered for resistance to pathogens and herbicides and for better nutrient profiles. GM livestock have been developed, although, as of November 2013, none were on the market.

There is a scientific consensus that currently available food derived from GM crops poses no greater risk to human health than conventional food, but that each GM food needs to be tested on a case-by-case basis before introduction. Nonetheless, members of the public are much less likely than scientists to perceive GM foods as safe. The legal and regulatory status of GM foods varies by country, with some nations banning or restricting them, and others permitting them with widely differing degrees of regulation.

However, there are ongoing public concerns related to food safety, regulation, labelling, environmental impact, research methods, and the fact that some GM seeds, along with all new plant varieties, are subject to plant breeders' rights owned by corporations.

Genetically modified foods are foods produced from organisms that have had changes introduced into their DNA using the methods of genetic engineering as opposed to traditional cross breeding. In the U.S., the Department of Agriculture (USDA) and the Food and Drug Administration (FDA) favor the use of the term genetic engineering over genetic modification as being more precise; the USDA defines genetic modification to include "genetic engineering or other more traditional methods".

According to the World Health Organization, "Foods produced from or using GM organisms are often referred to as GM foods."

Human-directed genetic manipulation of food began with the domestication of plants and animals through artificial selection at about 10,500 to 10,100 BC. The process of selective breeding, in which organisms with desired traits (and thus with the desired genes) are used to breed the next generation and organisms lacking the trait are not bred, is a precursor to the modern concept of genetic modification (GM). With the discovery of DNA in the early 1900s and various advancements in genetic techniques through the 1970s it became possible to directly alter the DNA and genes within food.

Genetically modified microbial enzymes were the first application of genetically modified organisms in food production and were approved in 1988 by the US Food and Drug Administration. In the early 1990s, recombinant chymosin was approved for use in several countries. Cheese had typically been made using the enzyme complex rennet that had been extracted from cows' stomach lining. Scientists modified bacteria to produce chymosin, which was also able to clot milk, resulting in cheese curds.

The first genetically modified food approved for release was the Flavr Savr tomato in 1994. Developed by Calgene, it was engineered to have a longer shelf life by inserting an antisense gene that delayed ripening. China was the first country to commercialize a transgenic crop in 1993 with the introduction of virus-resistant tobacco. In 1995, Bacillus thuringiensis (Bt) Potato was approved for cultivation, making it the first pesticide producing crop to be approved in the US. Other genetically modified crops receiving marketing approval in 1995 were: canola with modified oil composition, Bt maize, cotton resistant to the herbicide bromoxynil, Bt cotton, glyphosate-tolerant soybeans, virus-resistant squash, and another delayed ripening tomato.

With the creation of golden rice in 2000, scientists had genetically modified food to increase its nutrient value for the first time.

By 2010, 29 countries had planted commercialized biotech crops and a further 31 countries had granted regulatory approval for transgenic crops to be imported. The US was the leading country in the production of GM foods in 2011, with twenty-five GM crops having received regulatory approval. In 2015, 92% of corn, 94% of soybeans, and 94% of cotton produced in the US were genetically modified strains.

The first genetically modified animal to be approved for food use was AquAdvantage salmon in 2015. The salmon were transformed with a growth hormone-regulating gene from a Pacific Chinook salmon and a promoter from an ocean pout enabling it to grow year-round instead of only during spring and summer.

In April 2016, a white button mushroom (Agaricus bisporus) modified using the CRISPR technique received de facto approval in the United States, after the USDA said it would not have to go through the agency's regulatory process. The agency considers the mushroom exempt because the editing process did not involve the introduction of foreign DNA.

The most widely planted GMOs are designed to tolerate herbicides. By 2006 some weed populations had evolved to tolerate some of the same herbicides. Palmer amaranth is a weed that competes with cotton. A native of the southwestern US, it traveled east and was first found resistant to glyphosate in 2006, less than 10 years after GM cotton was introduced.

Process

Creating genetically modified food is a multi-step process. The first step is to identify a useful gene from another organism that you would like to add. The gene can be taken from a cell or artificially synthesised, and then combined with other genetic elements, including a promoter and terminator region and a selectable marker. Then the genetic elements are inserted into the targets genome. DNA is generally inserted into animal cells using microinjection, where it can be injected through the cell's nuclear envelope directly into the nucleus, or through the use of viral vectors. In plants the DNA is often inserted using Agrobacterium-mediated recombination, biolistics or electroporation. As only a single cell is transformed with genetic material, the organism must be regenerated from that single cell. In plants this is accomplished through tissue culture. In animals it is necessary to ensure that the inserted DNA is present in the embryonic stem cells. Further testing using PCR, Southern hybridization, and DNA sequencing is conducted to confirm that an organism contains the new gene.

Traditionally the new genetic material was inserted randomly within the host genome. Gene targeting techniques, which creates double-stranded breaks and takes advantage on the cells natural homologous recombination repair systems, have been developed to target insertion to exact locations. Genome editing uses artificially engineered nucleases that create breaks at specific points. There are four families of engineered nucleases: meganucleases, zinc finger nucleases, transcription activator-like effector nucleases (TALENs), and the Cas9-guideRNA system (adapted from CRISPR). TALEN and CRISPR are the two most commonly used and each has its own advantages. TALENs have greater target specificity, while CRISPR is easier to design and more efficient.

Crops

Genetically modified crops (GM crops) are genetically modified plants that are used in agriculture. The first crops developed were used for animal or human food and provide resistance to certain pests, diseases, environmental conditions, spoilage or chemical treatments (e.g. resistance to a herbicide). The second generation of crops aimed to improve the quality, often by altering the nutrient profile. Third generation genetically modified crops could be used for non-food purposes, including the production of pharmaceutical agents, biofuels, and other industrially useful goods, as well as for bioremediation. GM crops have been produced to improve harvests through reducing insect pressure, increase nutrient value and tolerate different abiotic stresses. As of 2018, the commercialised crops are limited mostly to cash crops like cotton, soybean, maize and canola and the vast majority of the introduced traits provide either herbicide tolerance or insect resistance.

The majority of GM crops have been modified to be resistant to selected herbicides, usually a glyphosate or glufosinate based one. Genetically modified crops engineered to resist herbicides are now more available than conventionally bred resistant varieties. Most currently available genes used to engineer insect resistance come from the Bacillus thuringiensis (Bt) bacterium and code for delta endotoxins. A few use the genes that encode for vegetative insecticidal proteins. The only gene commercially used to provide insect protection that does not originate from B. thuringiensis is the Cowpea trypsin inhibitor (CpTI). CpTI was first approved for use cotton in 1999 and is currently undergoing trials in rice. Less than one percent of GM crops contained other traits, which include providing virus resistance, delaying senescence and altering the plants composition.

Adoption by farmers has been rapid, between 1996 and 2013, the total surface area of land cultivated with GM crops increased by a factor of 100. Geographically though the spread has been uneven, with strong growth in the Americas and parts of Asia and little in Europe and Africa. Its socioeconomic spread has been more even, with approximately 54% of worldwide GM crops grown in developing countries in 2013. Although doubts have been raised, most studies have found growing GM crops to be beneficial to farmers through decreased pesticide use as well as increased crop yield and farm profit.

Fruits and Vegetables

Three views of a papaya, cultivar "Sunset", which was genetically modified to create the cultivar 'SunUp', which is resistant to Papaya ringspot virus.

Papaya was genetically modified to resist the ringspot virus (PSRV). "SunUp" is a transgenic red-fleshed Sunset papaya cultivar that is homozygous for the coat protein gene PRSV; "Rainbow" is a yellow-fleshed F1 hybrid developed by crossing 'SunUp' and nontransgenic yellow-fleshed "Kapoho". The GM cultivar was approved in 1998 and by 2010 80% of Hawaiian papaya was genetically engineered. The New York Times stated, "without it, the state's papaya industry would have collapsed". In China, a transgenic PRSV-resistant papaya was developed by South China Agricultural University and was first approved for commercial planting in 2006; as of 2012 95% of the papaya grown in Guangdong province and 40% of the papaya grown in Hainan province was genetically modified. In Hong Kong, where there is an exemption on growing and releasing any varieties of GM papaya, more than 80% of grown and imported papayas were transgenic.

The New Leaf potato, a GM food developed using Bacillus thuringiensis (Bt), was made to provide in-plant protection from the yield-robbing Colorado potato beetle. The New Leaf potato, brought to market by Monsanto in the late 1990s, was developed for the fast food market. It was withdrawn in 2001 after retailers rejected it and food processors ran into export problems.In 2011, BASF requested the European Food Safety Authority's approval for cultivation and marketing of its Fortuna potato as feed and food. The potato was made resistant to late blight by adding resistant genes blb1 and blb2 that originate from the Mexican wild potato Solanum bulbocastanum. In February 2013, BASF withdrew its application. In 2014, the USDA approved a genetically modified potato developed by J. R. Simplot Company that contained ten genetic modifications that prevent bruising and produce less acrylamide when fried. The modifications eliminate specific proteins from the potatoes, via RNA interference, rather than introducing novel proteins.

As of 2005, about 13% of the Zucchini (a form of squash) grown in the US was genetically modified to resist three viruses; that strain is also grown in Canada.

Plums genetically engineered for resistance to plum pox,
a disease carried by aphids.

In 2013, the USDA approved the import of a GM pineapple that is pink in color and that "overexpresses" a gene derived from tangerines and suppress other genes, increasing production of lycopene. The plant's flowering cycle was changed to provide for more uniform growth and quality. The fruit "does not have the ability to propagate and persist in the environment once they have been harvested", according to USDA APHIS. According to Del Monte's submission, the pineapples are commercially grown in a "monoculture" that prevents seed production, as the plant's flowers aren't exposed to compatible pollen sources. Importation into Hawaii is banned for "plant sanitation" reasons.

In February 2015 Arctic Apples were approved by the USDA, becoming the first genetically modified apple approved for sale in the US. Gene silencing is used to reduce the expression of polyphenol oxidase (PPO), thus preventing the fruit from browning.

Corn

Corn used for food and ethanol has been genetically modified to tolerate various herbicides and to express a protein from Bacillus thuringiensis (Bt) that kills certain insects. About 90% of the corn grown in the US was genetically modified in 2010. In the US in 2015, 81% of corn acreage contained the Bt trait and 89% of corn acreage contained the glyphosate-tolerant trait. Corn can be processed into grits, meal and flour as an ingredient in pancakes, muffins, doughnuts, breadings and batters, as well as baby foods, meat products, cereals and some fermented products. Corn-based masa flour and masa dough are used in the production of taco shells, corn chips and tortillas.

Soy

Soybeans accounted for half of all genetically modified crops planted in 2014. Genetically modified soybean has been modified to tolerate herbicides and produce healthier oils. In 2015, 94% of soybean acreage in the U.S. was genetically modified to be glyphosate-tolerant.

Rice

Golden rice is the most well-known GM crop that is aimed at increasing nutrient value. It has been engineered with three genes that biosynthesise beta-carotene, a precursor of vitamin A, in the edible parts of rice. It is intended to produce a fortified food to be grown and consumed in areas with a shortage of dietary vitamin A, a deficiency which each year is estimated to kill 670,000 children under the age of 5 and cause an additional 500,000 cases of irreversible childhood blindness. The original golden rice produced 1.6μg/g of the carotenoids, with further development increasing this 23 times. In 2018 it gained its first approvals for use as food.

Wheat

As of December 2017, genetically modified wheat has been evaluated in field trials, but has not been released commercially.

Derivative Products

Corn Starch and Starch Sugars Including Syrups

Starch or amylum is a polysaccharide produced by all green plants as an energy store. Pure starch is a white, tasteless and odourless powder. It consists of two types of molecules: the linear and helical amylose and the branched amylopectin. Depending on the plant, starch generally contains 20 to 25% amylose and 75 to 80% amylopectin by weight.

Starch can be further modified to create modified starch for specific purposes, including creation of many of the sugars in processed foods. They include:

- Maltodextrin, a lightly hydrolyzed starch product used as a bland-tasting filler and thickener.
- Various glucose syrups, also called corn syrups in the US, viscous solutions used as sweeteners and thickeners in many kinds of processed foods.
- Dextrose, commercial glucose, prepared by the complete hydrolysis of starch.
- High fructose syrup, made by treating dextrose solutions with the enzyme glucose isomerase, until a substantial fraction of the glucose has been converted to fructose.
- Sugar alcohols, such as maltitol, erythritol, sorbitol, mannitol and hydrogenated starch hydrolysate, are sweeteners made by reducing sugars.

Lecithin

Lecithin is a naturally occurring lipid. It can be found in egg yolks and oil-producing plants. It is an emulsifier and thus is used in many foods. Corn, soy and safflower oil are sources of lecithin, though the majority of lecithin commercially available is derived from soy. Sufficiently processed lecithin is often undetectable with standard testing practices. According to the FDA, no evidence shows or suggests hazard to the public when lecithin is used at common levels. Lecithin added to foods amounts to only 2 to 10 percent of the 1 to 5 g of phosphoglycerides consumed daily on average. Nonetheless, consumer concerns about GM food extend to such products. This concern led to policy and regulatory changes in Europe in 2000, when Regulation (EC) 50/2000 was passed which required labelling of food containing additives derived from GMOs, including lecithin. Because of the difficulty of detecting the origin of derivatives like lecithin with current testing practices, European regulations require those who wish to sell lecithin in Europe to employ a comprehensive system of Identity preservation (IP).

Sugar

The US imports 10% of its sugar, while the remaining 90% is extracted from sugar beet and sugarcane. After deregulation in 2005, glyphosate-resistant sugar beet was extensively adopted in the United States. 95% of beet acres in the US were planted with glyphosate-resistant seed in 2011. GM sugar beets are approved for cultivation in the US, Canada and Japan; the vast majority are grown in the US. GM beets are approved for import and consumption in Australia, Canada, Colombia, EU, Japan, Korea, Mexico, New Zealand, Philippines, the Russian Federation and Singapore. Pulp from the refining process is used as animal feed. The sugar produced from GM sugar beets contains no DNA or protein – it is just sucrose that is chemically indistinguishable from sugar produced from non-GM sugar beets. Independent analyses conducted by internationally recognized laboratories found that sugar from Roundup Ready sugar beets is identical to the sugar from comparably grown conventional (non-Roundup Ready) sugar beets.

Vegetable Oil

Most vegetable oil used in the US is produced from GM crops canola, corn, cotton and soybeans. Vegetable oil is sold directly to consumers as cooking oil, shortening and margarine and is used in prepared foods. There is a vanishingly small amount of protein or DNA from the original crop in vegetable oil. Vegetable oil is made of triglycerides extracted from plants or seeds and then refined and may be further processed via hydrogenation to turn liquid oils into solids. The refining process removes all, or nearly all non-triglyceride ingredients. Medium-chain triglycerides (MCTs) offer an alternative to conventional fats and oils. The length of a fatty acid influences its fat absorption during the digestive process. Fatty acids in the middle position on the glycerol molecules appear to be absorbed more easily and influence metabolism more than fatty acids on the end positions. Unlike ordinary fats, MCTs are metabolized like carbohydrates. They have exceptional oxidative stability, and prevent foods from turning rancid readily.

Other Uses

Animal Feed

Livestock and poultry are raised on animal feed, much of which is composed of the leftovers from processing crops, including GM crops. For example, approximately 43% of a canola seed is oil. What remains after oil extraction is a meal that becomes an ingredient in animal feed and contains canola protein. Likewise, the bulk of the soybean crop is grown for oil and meal. The high-protein defatted and toasted soy meal becomes livestock feed and dog food. 98% of the US soybean crop goes for livestock feed. In 2011, 49% of the US maize harvest was used for livestock feed (including the percentage of waste from distillers grains). "Despite methods that are becoming more and more sensitive, tests have not yet been able to establish a difference in the meat, milk, or eggs of animals depending on the type of feed they are fed. It is impossible to tell if an animal was fed GM soy just by looking at the resulting meat, dairy, or egg products. The only way to verify the presence of GMOs in animal feed is to analyze the origin of the feed itself."

A 2012 literature review of studies evaluating the effect of GM feed on the health of animals did not find evidence that animals were adversely affected, although small biological differences were occasionally found. The studies included in the review ranged from 90 days to two years, with several of the longer studies considering reproductive and intergenerational effects.

Enzymes produced by genetically modified microorganisms are also integrated into animal feed to enhance availability of nutrients and overall digestion. These enzymes may also provide benefit to the gut microbiome of an animal, as well as hydrolyse antinutritional factors present in the feed.

Proteins

Rennet is a mixture of enzymes used to coagulate milk into cheese. Originally it was available only from the fourth stomach of calves, and was scarce and expensive, or was available from microbial sources, which often produced unpleasant tastes. Genetic engineering made it possible to extract rennet-producing genes from animal stomachs and insert them into bacteria, fungi or yeasts to make them produce chymosin, the key enzyme. The modified microorganism is killed after fermentation. Chymosin is isolated from the fermentation broth, so that the Fermentation-Produced Chymosin (FPC) used by cheese producers has an amino acid sequence that is identical to bovine rennet. The majority of the applied chymosin is retained in the whey. Trace quantities of chymosin may remain in cheese.

FPC was the first artificially produced enzyme to be approved by the US Food and Drug Administration. FPC products have been on the market since 1990 and as of 2015 had yet to be surpassed in commercial markets. In 1999, about 60% of US hard cheese was made with FPC. Its global market share approached 80%. By 2008, approximately 80% to 90% of commercially made cheeses in the US and Britain were made using FPC.

In some countries, recombinant (GM) bovine somatotropin (also called rBST, or bovine growth hormone or BGH) is approved for administration to increase milk production. rBST may be present in milk from rBST treated cows, but it is destroyed in the digestive system and even if directly injected into the human bloodstream, has no observable effect on humans. The FDA, World Health Organization, American Medical Association, American Dietetic Association and the National Institutes of Health have independently stated that dairy products and meat from rBST-treated cows are safe for human consumption. However, on 30 September 2010, the United States Court of Appeals, Sixth Circuit, analyzing submitted evidence, found a "compositional difference" between milk from rBGH-treated cows and milk from untreated cows. The court stated that milk from rBGH-treated cows has: increased levels of the hormone Insulin-like growth factor 1 (IGF-1); higher fat content and lower protein content when produced at certain points in the cow's lactation cycle; and more somatic cell counts, which may "make the milk turn sour more quickly".

Livestock

Genetically modified livestock are organisms from the group of cattle, sheep, pigs, goats, birds, horses and fish kept for human consumption, whose genetic material (DNA) has been altered using genetic engineering techniques. In some cases, the aim is to introduce a new trait to the animals which does not occur naturally in the species, i.e. transgenesis.

A 2003 review published on behalf of Food Standards Australia New Zealand examined transgenic experimentation on terrestrial livestock species as well as aquatic species such as fish and shellfish. The review examined the molecular techniques used for experimentation as well as techniques for tracing the transgenes in animals and products as well as issues regarding transgene stability.

Some mammals typically used for food production have been modified to produce non-food products, a practice sometimes called Pharming.

Salmon

A GM salmon, awaiting regulatory approval since 1997, was approved for human consumption by the American FDA in November 2015, to be raised in specific land-based hatcheries in Canada and Panama.

Health and Safety

There is a scientific consensus that currently available food derived from GM crops poses no greater risk to human health than conventional food, but that each GM food needs to be tested on a case-by-case basis before introduction. Nonetheless, members of the public are much less likely than scientists to perceive GM foods as safe. The legal and regulatory status of GM foods varies by country, with some nations banning or restricting them, and others permitting them with widely differing degrees of regulation.

Opponents claim that long-term health risks have not been adequately assessed and propose various combinations of additional testing, labeling or removal from the market. The advocacy group European Network of Scientists for Social and Environmental Responsibility (ENSSER), disputes the claim that "science" supports the safety of current GM foods, proposing that each GM food must be judged on case-by-case basis.

Testing

The legal and regulatory status of GM foods varies by country, with some nations banning or restricting them, and others permitting them with widely differing degrees of regulation. Countries such as the United States, Canada, Lebanon and Egypt use substantial equivalence to determine if further testing is required, while many countries such as those in the European Union, Brazil and China only authorize GMO cultivation on a case-by-case basis. In the U.S. the FDA determined that GMO's are "Generally Recognized as Safe" (GRAS) and therefore do not require additional testing if the GMO product is substantially equivalent to the non-modified product. If new substances are found, further testing may be required to satisfy concerns over potential toxicity, allergenicity, possible gene transfer to humans or genetic outcrossing to other organisms.

Regulation

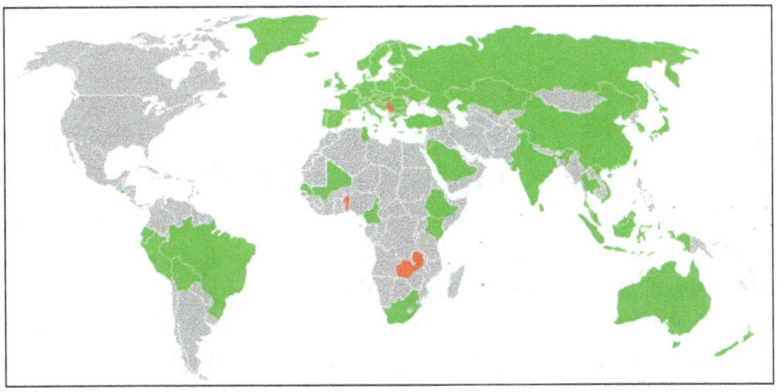

Green: Mandatory labeling required; Red: Ban on import and cultivation of genetically engineered food.

Government regulation of GMO development and release varies widely between countries. Marked differences separate GMO regulation in the U.S. and GMO regulation in the European Union. Regulation also varies depending on the intended product's use. For example, a crop not intended for food use is generally not reviewed by authorities responsible for food safety.

United States Regulations

In the U.S., three government organizations regulate GMOs. The FDA checks the chemical composition of organisms for potential allergens. The United States Department of Agriculture (USDA) supervises field testing and monitors the distribution of GM seeds. The United States Environmental Protection Agency (EPA) is responsible for monitoring pesticide usage, including plants modified to contain proteins toxic to insects. Like USDA, EPA also oversees field testing and the distribution of crops that have had contact with pesticides to ensure environmental safety. In 2015 the Obama administration announced that it would update the way the government regulated GM crops.

In 1992 FDA published "Statement of Policy: Foods derived from New Plant Varieties". This statement is a clarification of FDA's interpretation of the Food, Drug, and Cosmetic Act with respect to foods produced from new plant varieties developed using recombinant deoxyribonucleic acid (rDNA) technology. FDA encouraged developers to consult with the FDA regarding any bioengineered foods in development. The FDA says developers routinely do reach out for consultations. In 1996 FDA updated consultation procedures.

The StarLink corn recalls occurred in the autumn of 2000, when over 300 food products were found to contain a genetically modified corn that had not been approved for human consumption. It was the first-ever recall of a genetically modified food.

Labeling

As of 2015, 64 countries require labeling of GMO products in the marketplace.

US and Canadian national policy is to require a label only given significant composition differences or documented health impacts, although some individual US states (Vermont, Connecticut and Maine) enacted laws requiring them. In July 2016, Public Law 114-214 was enacted to regulate labeling of GMO food on a national basis.

In some jurisdictions, the labeling requirement depends on the relative quantity of GMO in the product. A study that investigated voluntary labeling in South Africa found that 31% of products labeled as GMO-free had a GM content above 1.0%.

In the European Union all food (including processed food) or feed that contains greater than 0.9% GMOs must be labelled.

Detection

Testing on GMOs in food and feed is routinely done using molecular techniques such as PCR and bioinformatics.

In a January 2010 paper, the extraction and detection of DNA along a complete industrial soybean oil processing chain was described to monitor the presence of Roundup Ready (RR) soybean: "The amplification of soybean lectin gene by end-point polymerase chain reaction (PCR) was successfully achieved in all the steps of extraction and refining processes, until the fully refined soybean oil. The amplification of RR soybean by PCR assays using event-specific primers was also achieved for all the extraction and refining steps, except for the intermediate steps of refining (neutralisation, washing and bleaching) possibly due to sample instability. The real-time PCR assays using specific probes confirmed all the results and proved that it is possible to detect and quantify genetically modified organisms in the fully refined soybean oil. To our knowledge, this has never been reported before and represents an important accomplishment regarding the traceability of genetically modified organisms in refined oils."

According to Thomas Redick, detection and prevention of cross-pollination is possible through the suggestions offered by the Farm Service Agency (FSA) and Natural Resources Conservation Service (NRCS). Suggestions include educating farmers on the importance of coexistence, providing farmers with tools and incentives to promote coexistence, conduct research to understand and monitor gene flow, provide assurance of quality and diversity in crops, provide compensation for actual economic losses for farmers.

SOMATIC FUSION

Somatic fusion, also called protoplast fusion, is a type of genetic modification in plants by which two distinct species of plants are fused together to form a new hybrid plant with the characteristics of both, a somatic hybrid. Hybrids have been produced either between different varieties of the same species (e.g. between non-flowering potato plants and flowering potato plants) or between two different species (e.g. between wheat Triticum and rye Secale to produce Triticale).

Uses of somatic fusion include making potato plants resistant to potato leaf roll disease. Through somatic fusion, the crop potato plant Solanum tuberosum – the yield of which is severely reduced by a viral disease transmitted on by the aphid vector – is fused with the wild, non-tuber-bearing potato Solanum brevidens, which is resistant to the disease. The resulting hybrid has the chromosomes of both plants and is thus similar to polyploid plants. Somatic hybridization was first introduced by Carlson et al. in Nicotiana glauca.

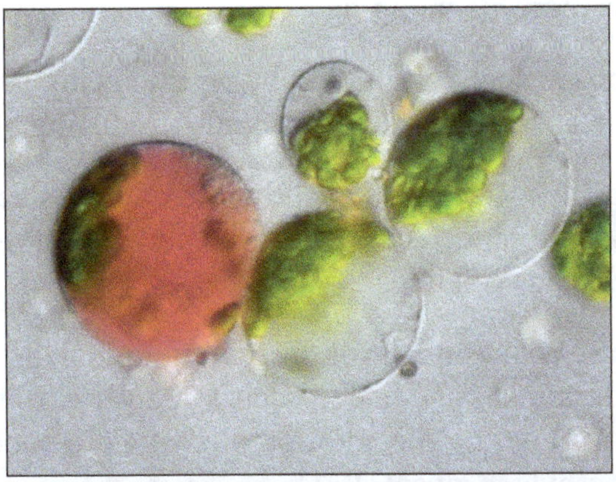

Fused protoplast (left) with chloroplasts (from a leaf cell) and coloured vacuole (from a petal).

Process for Plant Cells

The somatic fusion process occurs in four steps:

- The removal of the cell wall of one cell of each type of plant using cellulase enzyme to produce a somatic cell called a protoplast.
- The cells are then fused using electric shock (electrofusion) or chemical treatment to join the cells and fuse together the nuclei. The resulting fused nucleus is called heterokaryon.
- The formation of the cell wall is then induced using hormones.
- The cells are then grown into calluses which then are further grown to plantlets and finally to a full plant, known as a somatic hybrid.

The procedure for seed plants describe above, fusion of moss protoplasts can be initiated without electric shock but by the use of polyethylene glycol (PEG). Further, moss protoplasts do not need phytohormones for regeneration, and they do not form a callus. Instead, regenerating moss protoplasts behave like germinating moss spores. Of further note sodium nitrate and calcium ion at high pH can be used, although results are variable depending on the organism.

Applications of Hybrid Cells

Somatic cells of different types can be fused to obtain hybrid cells. Hybrid cells are useful in a variety of ways:

- To study the control of cell division and gene expression.
- To investigate malignant transformations.
- To obtain viral replication.
- For gene or chromosome mapping.
- Production of monoclonal antibodies by producing hybridoma (hybrid cells between an immortalised cell and an antibody producing lymphocyte), etc.

Chromosome mapping through somatic cell hybridization is essentially based on fusion of human and mouse somatic cells. Generally, human fibrocytes or leucocytes are fused with mouse continuous cell lines.

When human and mouse cells (or cells of any two mammalian species or of the same species) are mixed, spontaneous cell fusion occurs at a very low rate (10-6). Cell fusion is enhanced 100 to 1000 times by the addition of ultraviolet inactivated Sendai (parainfluenza) virus or polyethylene glycol (PEG).

These agents adhere to the plasma membranes of cells and alter their properties in such a way that facilitates their fusion. Fusion of two cells produces a heterokaryon, i.e., a single hybrid cell with two nuclei, one from each of the cells entering fusion. Subsequently, the two nuclei also fuse to yield a hybrid cell with a single nucleus.

A generalized scheme for somatic cell hybridization may be described as follows. Appropriate human and mouse cells are selected and mixed together in the presence of inactivated Sendai virus or PEG to promote cell fusion. After a period of time, the cells (a mixture of man, mouse and 'hybrid' cells) are plated on a selective medium, e.g., HAT medium, which allows the multiplication of hybrid cells only.

Several clones (each derived from a single hybrid cell) of the hybrid cells are thus isolated and subjected to both cytogenetic and appropriate biochemical analyses for the detection of enzyme/ protein/trait under investigation. An attempt is now made to correlate the presence and absence of the trait with the presence and absence of a human chromosome in the hybrid clones.

If there is a perfect correlation between the presence and absence of a human chromosome and that of a trait in the hybrid clones, the gene governing the trait is taken to be located in the concerned chromosome.

The HAT medium is one of the several selective media used for the selection of hybrid cells. This medium is supplemented with hypoxanthine, aminopterin and thymidine, hence the name HAT medium. Antimetabolite aminopterin blocks the cellular biosynthesis of purines and pyrimidines from simple sugars and amino acids.

However, normal human and mouse cells can still multiply as they can utilize hypoxanthine and thymidine present in the medium through a salvage pathway, which ordinarily recycles the purines and pyrimidines produced from degradation of nucleic acids.

Hypoxanthine is converted into guanine by the enzyme hypoxanthine-guanine phosphoribosyltransferase (HGPRT), while thymidine is phosphorylated by thymidine kinase (TK); both HGPRT and TK are enzymes of the salvage pathway.

On a HAT medium, only those cells that have active HGPRT (HGPRT+) and TK (TK+) enzymes can proliferate, while those deficient in these enzymes (HGPRr- and/or TK-) can not divide (since they cannot produce purines and pyrimidines due to the aminopterin present in the HAT medium).

For using HAT medium as a selective agent, human cells used for fusion must be deficient for either the enzyme HGPRT or TK, while mouse cells must be deficient for the other enzyme of this pair. Thus, one may fuse HGPRT deficient human cells (designated as TK+ HGPRr-) with TK deficient mouse cells (denoted as TK- HGPRT+).

Their fusion products (hybrid cells) will be TK+ (due to the human gene) and HGPRT+ (due to the mouse gene) and will multiply on the HAT medium, while the man and mouse cells will fail to do so. Experiments with other selective media can be planned in a similar fashion.

Characteristics of Somatic Hybridization and Cybridization

- Somatic cell fusion appears to be the only means through which two different parental genomes can be recombined among plants that cannot reproduce sexually (asexual or sterile).
- Protoplasts of sexually sterile (haploid, triploid, and aneuploid) plants can be fused to produce fertile diploids and polyploids.
- Somatic cell fusion overcomes sexual incompatibility barriers. In some cases somatic hybrids between two incompatible plants have also found application in industry or agriculture.

- Somatic cell fusion is useful in the study of cytoplasmic genes and their activities and this information can be applied in plant breeding experiments.

Inter-specific and Inter-generic Fusion Achievements

The table only lists a few examples, there are many more crosses. The possibilities of this technology are great; however, not all species are easily put into protoplast culture.

Cross	Crossed with
Oat	Maize
Brassica sinensis	B. oleracea
Torrentia fourneri	T. bailloni
Brassica oleracea	B. campestris
Datura innoxia	Atropa belladonna
Nicotiana tabacum	N. glutinosa
Datura innoxia	D. candida
Arabidopsis thaliana	Brassica campestris
Petunia hybrida	Vicia faba

TRANSGENIC PLANTS

Transgenic plants are plants that have been genetically modified by inserting genes directly into a single plant cell. Transgenic crop plants modified for improved flavor, pest resistance, or some other useful property are being used increasingly.

Transgenic plants are unique in that they develop from only one plant cell. In normal sexual reproduction, plant offspring are created when a pollen cell and an ovule fuse. In a similar laboratory procedure, two plant cells that have had their cell walls removed can be fused to create an offspring.

There are three general approaches that can be used to insert the DNA into a plant cell: vector-mediated transformation, particle-mediated transformation, and direct DNA absorption. With vector-mediated transformation, a plant cell is infected with a virus or bacterium that, as part of the infection process, inserts the DNA. The most commonly used vector is the crown-gall bacterium, Agrobacterium tumefaciens. With particle-mediated transformation (particle bombardment), using a tool referred to as a "gene gun," the DNA is carried into the cell by metal particles that have been accelerated, A "normal" melon (left) after five days, and a transgenic melon (right) after fifteen days. The ability of the modified fruit to be edible for a much longer period is a trait desirable for those in the produce business or "shot," into the cell. The particles are usually very fine gold pellets onto which the DNA has been stuck. With direct DNA absorption, a cell is bathed in the DNA, and an electric shock usually is applied ("electroporation") to the cell to stimulate DNA uptake.

No matter what gene insertion method is used, a series of events must occur to allow a whole genetically modified plant to be recovered from the genetically modified cell: The cell must incorporate the new DNA into its own chromosomes, the transformed cell must initiate division, the new cells need to organize themselves into all the tissues and organs of a normal plant ("regeneration"), and finally, the inserted gene must continue to work properly ("gene expression") in the regenerated plant.

To help ensure all this occurs, a "cassette" of genes is inserted during the initial transformation. In addition to the gene coding for the desired trait, other genes are added. Some of these genes promote the growth of only those plant cells that have successfully incorporated the inserted DNA. It might do this by providing the transformed cells with resistance to a normally toxic antibiotic that is added to the growth medium, for example. Other genes ("promoters") may be added to control the functioning of the trait gene by directing when and where in the transformed plant it will operate.

The genes put into plants using genetic engineering can come from any organism. Most genes used in the genetic engineering of plants have come from bacteria. However, as scientists learn more about the genetic makeup of plants ("plant genomics"), more plant-derived genes will be used.

Inserted genes can be classified into three groups based on their use: those that protect a crop, those that improve the quality of a harvested product, and those that let the plant perform some new function.

Genes that Protect a Crop

The major use of plant genetic engineering has been to make crops easier to grow by decreasing the impact of pests. Insect resistance has been achieved by transforming a crop using a Bt gene. Bt genes were isolated from Bacillus thuringiensis, a common soil bacterium. They code for proteins that severely disrupt the digestive system of insects. Thus an insect eating the leaf of a plant expressing a Bt gene stops eating and dies of starvation. There are many Bt genes, each of which targets a particular group of insects. Some Bt genes, for example, target caterpillars. Others target beetles.

Genetic engineering also has been used in the battle against weeds. Bacterial genes allow crops to either degrade herbicides or be resistant to them. The herbicides that are used are generally very effective, killing most plants. They are considered environmentally benign, degrading rapidly in the soil and having little impact on humans or other organisms. Thus a whole field of transgenic crops can be sprayed with broad-spectrum herbicides, killing all plants except the crops. Corn, soybeans, canola, and cotton that have been engineered to withstand either insects or herbicides, or both, are widely planted in some countries, including the United States. In addition, other crops, including potatoes, tomatoes, tropical fruits, and melons, have been engineered for resistance to viral diseases.

Genes that Improve Crop Quality

An emerging major use of genetic engineering for crops is to alter the quality of the crop. Fresh fruits and vegetables begin to deteriorate immediately after being harvested. Delaying or preventing this deterioration not only preserves a produce's flavor, and appearance, but maintains the nutritional value of the produce. Genes that change the hormonal status of the harvested crops are the major targets for genetic engineering toward longer shelf-life.

For example, the plant hormone ethylene is associated with accelerated ripening, as well as leaf and flower deterioration, in fruits that are injured or harvested. Scientists insert genes that interfere with a plant's ability to synthesize or respond to ethylene, thereby extending postharvest quality for many fresh products, including tomatoes, lettuce, and cut flowers. Scientists are also using gene insertion to improve a plant's nutritional value and color.

Genes that Introduce New Traits

One approach to improving the economic value of crops is to give them traits that are completely new for that plant. Some crops, including potatoes, tomatoes, and bananas, have been engineered with genes from pathogenic organisms. This is done to make animals, including humans that eat the crops immune to the diseases caused by the pathogens. The genes code for proteins that act as antigens to induce immunity. Thus edible parts of plants are engineered to act as oral vaccines. This approach may be particularly effective for pathogens, such as those causing diarrhea and other gastrointestinal disorders that enter the body through mucous membranes. This is because the "medicine" in the food comes into direct contact with these membranes and does not have to be absorbed into the blood stream. Genes have also been engineered into crop plants to direct the plants to produce industrial enzymes used in the manufacture of paper. Other genes direct crops to produce small polymers useful in the manufacture of plastics. This general approach is being termed "plant molecular farming."

Rice is another plant that has been engineered for a new trait. During commercial processing, a substantial part of the white rice grains are removed, leaving very little vitamin A. Vitamin A deficiency is a significant health problem in regions dependent on rice as a dietary staple. Scientists engineered a certain form of rice, known as "golden rice" because it has a yellow tinge, to express three introduced genes. These genes let the plant produce the precursor of vitamin A in the portion of the grain that remains after processing, thereby providing a dietary source of the vitamin.

References

- Genomics-in-agriculture-327499811: researchgate.net, Retrieved 18 April, 2020
- Genetic-resources-in-crop-production-biotechnology, biotechnology-genetically-modified-899: biotechnologynotes.com, Retrieved 28 July, 2020
- Qaim, Matin (2016-04-29). "Introduction". Genetically Modified Crops and Agricultural Development. Springer. pp. 1–10. ISBN 9781137405722
- Herbicide-tolerance, programs-genetic-engineering: beyondpesticides.org, Retrieved 10 March, 2020
- Transgenic-Plants-2902: medicine.jrank.org, Retrieved 17 June, 2020
- Demont, M.; Stein, A. J. (2013). "Global value of GM rice: A review of expected agronomic and consumer benefits". New Biotechnology. 30 (5): 426–436. doi:10.1016/j.nbt.2013.04.004. PMID 23628812
- Hamill, John D.; Cocking, Edward C. "Somatic Hybridization of Plants and its Use in Agriculture". Plant Cell Biotechnology. 18. doi:10.1007/978-3-642-73157-0_3

5

Biotechnological Approaches for Pest Management

Biotechnology in pest management is the field of science that targets insect-pest control via detailed analysis and engineering of biological processes. Some of the approaches include usage of transgenic plants, gene-encoding toxins, protein inhibitors, etc. This chapter sheds light on various biotechnological approaches for pest management to provide a thorough understanding of the subject.

Insect-resistant crops have been one of the major successes of applying plant genetic engineering technology to agriculture; cotton (Gossypium hirsutum) resistant to lepidopteran larvae (caterpillars) and maize (Zea mays) resistant to both lepidopteran and coleopteran larvae (rootworms) have become widely used in global agriculture and have led to reductions in pesticide usage and lower production costs.

The source of the insecticidal toxins produced in commercial transgenic plants is the soil bacterium Bacillus thuringiensis (Bt). Bt strains show differing specificities of insecticidal activity toward pests, and constitute a large reservoir of genes encoding insecticidal proteins, which are accumulated in the crystalline inclusion bodies produced by the bacterium on sporulation (Cry proteins, Cyt proteins) or expressed during bacterial growth (Vip proteins). The three-domain Cry proteins have been extensively studied; their mechanism of action involves a proteolytic activation step, which occurs in the insect gut after ingestion, followed by interaction of one or both of domains II and III with "receptors" on the surface of cells of the insect gut epithelium. This interaction leads to oligomerization of the protein, and domain I is then responsible for the formation of an open channel through the cell membrane. The resulting ionic leakage destroys the cell, leading to breakdown of the gut, bacterial proliferation, and insect death.

However, not all pests are adequately targeted by the Bt toxins used at present, and there is still a need to develop solutions to specific problems, such as resistance to sap-sucking pests and pests of stored products.

Plastid Genome Transformation

Expression of Bt toxins in transgenic plants needs to be at a sufficient level to confer adequate protection against target pests (defined by the Environmental Protection

Agency as giving >95% mortality of insects heterozygous for a resistance allele; in practice, generally >0.2% of total soluble protein in the appropriate tissue). Transformation of the nuclear genome with genes encoding Bt toxins gives very low levels of expression unless extensive modifications, which include removal of AT-rich regions from the coding sequence and use of modified constitutive or tissue-specific promoters, are carried out. These methods were established within the first stage of the development of this technology and are now considered routine, although they do pose significant technical problems.

In contrast, introduction of unmodified Bt genes into the chloroplast genome results in high levels of toxin accumulation, as the plastid genome is bacterial in origin. This method has not been widely adopted, due to significant technical problems in achieving stable transformation of the plastid genome and in transforming plastids in species other than tobacco (Nicotiana tabacum). Nevertheless, these difficulties can be overcome; various Cry proteins have been expressed in plastids of tobacco, and Cry1Ab has been expressed in soybean (Glycine max) plastids. Overexpression of the Cry2Aa2 operon in plastids is effective in giving broad-spectrum protection against a range of pests. This technique has the potential advantage that plastid-encoded characteristics are predominantly maternally inherited in most plants, so that pollen from transgenic crops is less likely to disperse the transgene to nontransgenic plant stocks. The inherent containment addresses concerns about the coexistence of transgenic and organic agricultural practices and may lead to this technique being more widely adopted for production of agricultural crops.

Novel Bt Toxins

Several novel Bt insecticidal proteins, which have no sequence similarity to three-domain Cry proteins, have been expressed in transgenic plants. Binary toxins require two components for activity and are exemplified by the Cry34/35 and Vip1/2 toxins, which are active against corn rootworm (Diabrotica virgifera). Cry34/35 have been expressed in transgenic maize and are the basis of a successful commercial product; no re-engineering of coding sequences was necessary, in contrast to an alternative approach using modified Cry3Bb1. Single-chain Vip proteins, such as Vip3, are active against lepidopteran larvae, with a broader range of toxicity than individual Cry proteins. Their range of activity can be further extended by protein engineering. Commercial transgenic plants expressing these proteins are under development.

Plants Expressing Multiple Toxins

The specificity of Bt Cry toxins toward target pest species is a major advantage in agriculture because effects on non-target insects and other organisms in the ecosystem are minimized. However, deployment of transgenic crops expressing a single specific Bt toxin can lead to problems in the field, where secondary pest species are not affected, and can cause significant damage to the crop. Introduction of additional Bt cry genes into the crop can afford protection against a wider range of pests. Commercial use of transgenic cotton containing two Bt genes began in 1999, 3 years after the release of the original single Bt variety. Cotton plants expressing both Cry1Ac and Cry2Ab proteins were more toxic to bollworms (Helicoverpa zea; target pest) and two species of armyworms (Spodoptera frugiperda and Spodoptera exigua; secondary pests) than cotton expressing Cry1Ac alone in laboratory trials and in greenhouse and field trials.

Expression of multiple Cry proteins can also be beneficial in prevention of resistance to toxin activity in the target pest(s). Although the "approved" refuge strategy has been highly successful in containing pest resistance to Bt toxins expressed in transgenic plants, targeting different receptors in the insect is (theoretically) more effective because multiple mutations are required to produce the loss of sensitivity to the toxins. Transgenic broccoli plants expressing either Cry1Ac or Cry1C or both proteins were exposed to a population of diamondback moth (Plutella xylostella), which carried Bt resistance genes at a relatively low frequency. After 24 pest generations, plants carrying two transgenes showed a significant delay in the selection of a resistant population of insects. However, simultaneous exposure of the insects to plants carrying either one or two Bt genes actually increased resistance breakdown in the two gene plants, suggesting that the multiple-toxin approach is not a cure-all solution. Other results have also shown that pests can acquire resistance to multiple toxins; for example, a strain of the lepidopteran cotton pest Heliothis virescens has simultaneous resistance to Cry1Ac and Cry2Aa, with a different genetic basis of resistance to each toxin.

Improvements in plant transformation methods, such as extending the species range of Agrobacterium-mediated gene transfer methods to monocots and using plasmid vectors containing multiple gene constructs to allow introduction of multiple transgenes at a single genetic locus, have enabled the expression of multiple toxins in transgenic plant varieties. The recent announcement of a transgenic maize variety containing six insect resistance genes active against corn rootworm and lepidopteran pests (rootworm; Cry34Ab1 + Cry35Ab1, modified Cry3Bb1: lepidoptera; Cry1F, Cry1A.105, Cry2Ab2) and two genes giving tolerance to herbicides (glyphosate and glufosinate-ammonium), as a "one-stop" solution to pest and weed problems, exemplifies this gene stacking technology.

Improving on Nature: Protein Engineering in Bt Toxins

Mutagenesis of Cry toxins has been used extensively in studying the mechanism of action of these proteins and has been exploited to produce novel recombinant toxins.

Domain Exchange in Three-domain Cry Toxins

The structural similarity of all members of the family of three-domain Bt toxins, and the separate roles of the domains in the processes of receptor binding and channel formation, suggested that combining domains from different proteins could generate active toxins with novel specificities. Transfer of the carbohydrate-binding domain III generated a Cry1Ab-Cry1C hybrid that was highly toxic to armyworm (S. exigua), an insect resistant to Cry1A toxins; the presence of the Cry1Ca domain III was sufficient to confer toxicity toward Spodoptera. More remarkably, a hybrid Cry protein, containing domains I and III from Cry1Ba and domain II of Cry1Ia, conferred resistance to the lepidopteran pest potato tuber moth (Phthorimaea operculella) and to the coleopteran Colorado potato beetle (Leptinotarsa decemlineata) when expressed in transgenic potato. The "parental" Cry proteins in this hybrid are lepidopteran specific, with no toxicity toward coleopterans such as the potato beetle, demonstrating the creation of a novel specificity.

Mutagenesis of Three-domain Cry Toxins

Modification of Bt toxins by site-directed mutagenesis to increase toxicity toward target pests has been employed as an alternative to the "domain swap" approach. The key role of domain II in three-domain Cry proteins in mediating interactions with insect receptors has been exploited by mutation of amino acid residues in the loop regions of this domain. Mutation of Cry1Ab increased its toxicity toward larvae of gypsy moth (Lymantria dispar) by up to 40-fold, and similar strategies were used to increase the toxicity of Cry3A protein toward target coleopteran pests. The level to which rational design of toxins is possible is shown by the engineering of toxicity toward mosquito larvae into the lepidopteran-specific toxin Cry1Aa. Alternatively, a directed evolution system based on phage display technology for producing toxins with improved binding to a receptor, and thus increased toxicity, has been described.

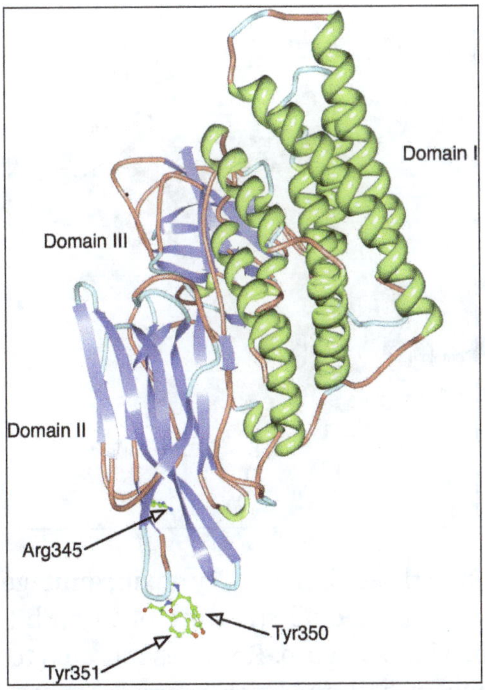

Engineering specificity in a three-domain Cry toxin; mutagenesis of the toxin-receptor interaction loop in domain II. Three-dimensional structure of Cry3A (1dlc; RCSB) is shown in ribbon format. Domain I (helices) is at top right, and domain II (sheet structure) is at bottom left. Domain III (carbohydrate-binding domain; sheet structure) is behind the other domains, central in this view. Residues mutated to increase toxicity toward yellow mealworm (Tenebrio molitor), Colorado potato beetle, and cottonwood leaf beetle (Chrysomela scripta) are shown in ball-and stick representation.

A current commercial transgenic maize variety with resistance to corn rootworm, MON863, expresses a modified version of the Bt Cry3Bb1 toxin. Unmodified Cry3Bb1 is active against a number of coleopteran species, but toxicity toward Western corn rootworm was not sufficient to give adequate protection at levels of expression achievable in maize. A large number of variants of the native Cry3Bb, incorporating a series of specific mutations that aimed to improve the channel-forming ability of the toxin, were produced and screened for activity. Mutations were carried out to: (1) increase hydrophobicity of the protein in regions in domain I containing sheets of bound water molecules and in loop regions; (2) increase the mobility of the channel-forming helices in domain I by disrupting hydrogen-bond formation; (3) increase the mobility and flexibility of loop regions in domain I; (4) alter potential ion-pair interactions and metal-binding sites; and (5) reduce or eliminate binding to carbohydrates in the insect gut by mutation of a loop region between domains I and II. The toxicity of the protein toward corn rootworm was increased approximately 8-fold, giving a product that could be expressed at levels sufficient for adequate protection against rootworm.

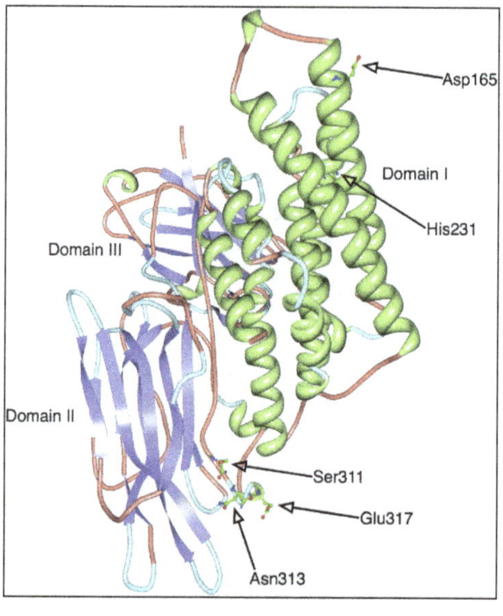

Engineering specificity in a three-domain Cry toxin; mutagenesis to improve channel-forming ability. Three-dimensional structure of Cry3Bb (1ji6; RCSB) is shown in ribbon format in the same view as figure. Residues mutated to increase toxicity toward corn rootworm are shown in ball-and-stick representation. Mutations are made in helices of domain I and in the region linking domains I and II. The mutation sites shown are taken from the most active toxin produced; a range of other sites for mutation were explored.

Recent results, showing that oligomerization of Cry toxins subsequent to binding to the cadherin "receptor" on the insect gut surface is a necessary step in the mechanism of toxicity, have led to a strategy to engineer Cry proteins to be effective against insects that have become resistant to normal toxins by receptor mutation. Removal of the α-1 helix of domain I resulted in a protein that did not require to bind to cadherin to oligomerize and was toxic to resistant insects. Expression of these modified toxins in plants has yet to be attempted.

Fusion Proteins

Transformation of plants with a gene construct containing a single translationally fused coding sequence encoding two Cry proteins has been used as an alternative to separate constructs, although there is no apparent advantage over simpler methods for introducing two genes. However, addition of sequences from other proteins can be used to introduce extra functionality into Cry toxins. For example, the Gal-binding lectin domain (B-chain) from the ribosome-inactivating protein ricin was fused C terminally to domain III of Cry1Ac, giving a fusion protein that could bind to Gal residues in potential receptors in the insect, in addition to binding to N-acetyl galactosamine residues by domain III of the toxin. Expression of the fusion protein in transgenic maize and rice (Oryza sativa) plants gave resistance to larvae of stemborers (Chilo suppressalis) and leaf armyworm (Spodoptera littoralis), whereas plants expressing the unmodified Cry1Ac were susceptible to both insects. Plants expressing the fusion protein were also resistant to a hemipteran pest, the leafhopper Cicadulina mbila; the lectin domain may cause this effect because Bt toxins are not effective against hemipterans.

Self Defense: Exploiting Plant Defensive Proteins

Engineering plants to express proteins that are end-products of the wounding response, such as proteinase inhibitors and polyphenol oxidase, has generally failed to give more than partial protection against insect herbivores, due to pre-adaptation by the pests. However, two examples of exploiting plant defensive proteins have shown promise in addressing specialized insect resistance problems.

Bean α-Amylase Inhibitors and Stored Product Pests

The α-amylase inhibitors from some legume seeds, which are similar to legume lectins in sequence, have been shown to be causative factors in the resistance of specific varieties of legumes to coleopteran seed weevils. The bean (Phaseolus vulgaris) α-amylase inhibitor gene was expressed in seeds of transgenic garden pea (Pisum sativum) and other grain legumes, using a strong seed-specific promoter. The resulting seeds contained up to 3% of the foreign protein and were resistant to stored product pests, such as larvae of bruchid beetles, and field pests, such as larvae of the pea weevil Bruchus pisorum. This strategy is directed toward coleopteran seed herbivores, which have a neutral or acidic gut pH, so the inhibitor is not inactivated, and in which starch digestion, not protein digestion, is nutritionally limiting.

Despite these results, agricultural deployment of transgenic crops expressing this α-amylase inhibitor gene has not taken place. Safety concerns have arisen as a result of the induction of systemic immunological responses in mice fed peas expressing the amylase inhibitor protein, which appear to result from altered posttranslational processing in pea compared to the "natural" host (bean).

Lectins and Sap-sucking Pests

Potential exploitation of lectin genes to confer insect resistance in transgenic plants has targeted hemipteran plant pests, which are not affected by known Bt toxins but have been shown to be susceptible to lectin toxicity. Expression of the Man-specific snowdrop lectin (GNA) in transgenic rice plants using constitutive or phloem-specific promoters gave plants that were partially resistant to rice brown planthopper (Nilaparvata lugens) and other hemipteran pests. Reductions of up to 50% in survival were observed, with reduced feeding, development, and fertility of survivors. Concerns about possible consequences to higher animals of ingesting snowdrop lectin have limited further progress, although a recent study incorporating a 90-d feeding trial found no adverse effects resulting from consumption of transgenic rice expressing GNA by rats. Similar partial resistance to hemipterans has also been obtained by expression of a Man-specific lectin from garlic (Allium sativum) leaves (ASA-L) in transgenic rice and a variety of other transgenic plant species. The transgenic rice plants expressing ASA-L were shown to decrease transmission of Rice tungro virus by its insect vector, presumably as a result of decreased feeding by the pest.

New Insecticidal Proteins

Photorhabdus Luminescens Insecticidal Proteins

Nematodes of Heterorhabditis species that contain symbiotic enterobacteria are widely used for small-scale biological control of insect pests. When nematodes enter an insect host, bacterial cells from the nematode gut are released into the insect circulatory system. Toxins secreted by the bacteria cause cell death in the insect host, leading to a lethal septicemia. P. luminescens, the most well-investigated bacterial species of this type, contains a large number of potentially insecticidal components. One of the orally toxic components, toxin A, was selected for further study. The encoding gene tcdA was cloned and assembled into expression constructs, containing 5′ and 3′ untranslated region sequences from a tobacco osmotin gene to improve expression levels of mRNA and protein in transgenic plants. Expression of toxin A at levels >0.07% of total soluble protein in leaves of transgenic Arabidopsis (Arabidopsis thaliana) plants gave almost complete protection against larvae of the lepidopteran tobacco hornworm (Manduca sexta). Leaf extracts from these plants were also toxic to corn rootworm, showing cross-species protection. Commercial development of this technique is likely.

Cholesterol Oxidase

Bacterial cholesterol oxidase has an insecticidal activity comparable to Bt toxins, dependent on its enzyme activity, which is thought to promote membrane destabilization. Expression constructs containing part or all of the coding sequence of the protein, or the coding sequence fused to a chloroplast-targeting peptide, resulted in production of active enzyme in transgenic tobacco. However, phenotypic abnormalities were observed in transgenic plants unless the enzyme was localized in chloroplasts, possibly as a result of interference with steroidal signaling pathways. Leaf tissue from all transgenic plants was toxic to boll weevil larvae. The cholesterol oxidase gene appears to be an obvious candidate for introduction into the chloroplast genome rather than the plant nuclear genome, which would avoid potential problems caused by enzyme activity in the cytoplasm.

Avidin as an Insecticidal Protein

Avidin has a strong insecticidal effect on many insects, although susceptibility varies widely between different insect species (apparently based on biotin requirements). Expression of avidin in transgenic maize initially aimed to produce the protein as a high-value product, but maize seed containing more than 0.1% avidin (of total protein) was fully resistant to larvae of three different coleopteran storage pests. The protein has also been expressed in other transgenic plants to confer pest resistance. Targeting of the foreign protein away from the cell cytoplasm is important to avoid developmental abnormalities in the plants. No further development of this promising method has been reported.

Exploiting Secondary Metabolism

Engineering Secondary Metabolism of Plant Defensive Compounds

The availability of genes encoding the biosynthetic enzymes of secondary metabolism has made transfer of biosynthetic pathways between plants feasible. Genes encoding two Cyt P450 oxidases and a UDP-glycosyltransferase from sorghum (Sorghum bicolor) have been transferred to Arabidopsis, resulting in the production of the cyanogenic glycoside dhurrin from Tyr. The resulting plants produced hydrogen cyanide on tissue damage and showed enhanced resistance to attack by the flea beetle Phyllotreta nemorum, a specialist feeder on crucifers. Other secondary metabolites that have been produced in transgenic plants include the alkaloid caffeine.

Engineering Secondary Metabolism of Volatile Communication Compounds

Engineering volatiles emitted by plants offers possibilities for new methods of crop protection. Volatile composition has been altered in tobacco by RNA interference (RNAi)-mediated suppression of a cytP450 oxidase gene expressed in trichomes, and in Arabidopsis by constitutive overexpression of a plastid dual linalool/nerolidol synthase. The transgenic plants deterred aphid colonization but were not wholly resistant. Volatiles can also be used as attractants for natural enemies of pests; for example, Arabidopsis plants transformed with the maize terpene synthase gene TPS10 emitted the sesquiterpene volatiles normally produced in maize and attracted parasitoid wasps that attack maize pests. Volatiles used by insects to communicate with each other can also be exploited; the sesquiterpene (E)-β-farnesene is an alarm pheromone in aphids and attracts aphid predators and parasitoids. When Arabidopsis was transformed with an (E)-β-farnesene synthase gene from mint (Mentha × piperita), the resulting plants showed significant levels of aphid deterrence in choice experiments and were attractive to the aphid parasitoid Diaeretiella rapae.

RNAi

Disrupting gene function by the use of RNAi is a well-established technique in insect genetics based on delivery by injection into insect cells or tissues. The observation that RNAi could also be effective in reducing gene expression, measured by mRNA level, when fed to insects has led to two recent articles in which transgenic plants producing double-stranded RNAs (dsRNAs) are shown to exhibit partial resistance to insect pests. Transgenic maize producing dsRNA directed against V-type ATPase of corn rootworm showed suppression of mRNA in the insect and reduction in feeding damage compared to controls. Similarly, transgenic tobacco and Arabidopsis plant material expressing dsRNA directed against a detoxification enzyme (Cyt P450 gene CYP6AE14) for gossypol in cotton bollworm caused the insect to become more sensitive to gossypol in the diet. This approach holds great promise for future development.

The production of transgenic insect-resistant plants, and their continuing development, has been a major scientific success, mirrored by the practical success of a limited number of pest-resistant genetically modified crops in some countries. However, these successes must be set against the failure to make this technology more widely available. In some developed countries, this has been a result of vocal opposition to plant genetic engineering technology itself; but in many instances, in both developed and developing countries, it is more a case of potential economic returns not being sufficient to make the introduction of engineered crop varieties commercially viable. There is a need to reconsider regulatory systems for release of transgenic crops. In most cases, these were set up with good intentions, but the end result in developed countries has been to make commercialization of transgenic crops difficult and expensive, so that only very large companies can afford to carry products through and then only when the projected returns are very large. Under these circumstances, it is not surprising that campaigns against genetically modified crops have been so successful, no matter how ill-founded their scientific basis may be.

In the developed world, the economic consequences to the public of the failure to adopt transgenic insect-resistant plants more widely have been seen as marginal, and the potential environmental benefits are discounted. The economic and environmental consequences for the developing world, where significant benefits to the lives of farmers could be obtained from this technology, are more serious. Here, access and implementation are being hindered by the actions of companies and organizations in developed countries. A change in attitude by governments, non-governmental organizations, and the public at large in some countries is needed for insect-resistant transgenic crops to be able to fulfill their promise to all the world's population, not just to the few.

USE OF BIOTECHNOLOGY IN THE CONTROL OF INSECTS-PRAGUE

Productivity gains in agriculture are satisfactory with the use of genetically modified plants and the dependency of application of insecticides on crops becomes smaller over the years. The consequences of the development and marketing of corn genetically modified (GM) have been profound, and in 2011 the area planted in the United States of America (USA) with at least one GM trait corresponded to more than 88% of was over acreage.

In addition, the efficiency gains in the production chains were only possible thanks to the entrepreneurship and management of rural producers, who adopted the most modern technologies available for science. Among these, stand: The tillage, fertilization and soil correction, the techniques of integrated management of invasive plants, diseases and insect pest and the growing adoption of improved seeds with high productive capacity. It is observed that the simple hybrids corn came to dominate the market of seeds embedded in technologies and seeds are more easily adopted by producers. This is the case of transgenic seeds, which in the culture of corn were widely adopted, including the major world producers of this cereal USA and Argentina.

In crops of corn, the losses caused by pests are limiting factor to achieve high productivity. The fall armyworm (Spodopterafrugiperda) is the main plague of corn in Brazil culture, causing severe damage. The attack on the plant occurs since its emergence to the booting and the silking. However, the critical period is flourishing. Losses due to Caterpillar attack may reduce production in up to 34%. Survey conducted among some 1,100 farmers who produce more than 8,000 kg ha-1 showed that, among the crops sampled, 15% received from 4 to 5 applications of insecticides and 6% received 6 to 8 applications for pest control.

Other recent surveys, conducted by Embrapa maize and sorghum, have shown that in some regions the number of applications of insecticides for the control of Caterpillar-cartridge can reach 10. In addition, there is no efficient method of chemical control for at least two other important species: Or the control of the corn earworm (Helicoverpazea) and the Maize stalk borer (Diatraeasaccharalis).

The insects have been one of the biggest causes of damage in food production being these losses of the order of 20 to 30% of world production. It is estimated that approximately 67,000 species of insects cause damage to plantations and tropical regions, usually the poorest in the world.

The attack of any pest depends on the development of culture, as well as the intensity of the attack, which can significantly affect the performance of the same. Chemical control is the primary measure used to prevent the immediate damage that reach the level of economic damage. Many times, the insecticides do not have the desired effectiveness and have a high cost, because are usually required multiple applications.

The insecticides used in pest control as an example in corn culture are often of low selectivity, therefore can affect the population of natural enemies, favoring the proliferation of pests and even resurgence of others. Due to these factors, the search for alternatives that can minimize or even replace the conventional insecticides was intensified and, currently, the new tactics comprise a series of alternatives: Resistant plants, selective insecticides, parasitoids and entomopathogenic microorganisms. Among the entomopathogenic Bacillus thuringiensis (Bt), notable for its wide use in the control of pest insects of the order Lepidoptera, especially in corn culture.

Currently, transgenic production is widespread in almost all agricultural regions of the planet and with the adoption of biotechnology by greater productivity with the producers reaches lower use of insecticides. An example of this is the use of gene technology (Cry) of bacteria (Bt) in control of the main pests of maize culture. The Bt gene technology diffusion aims to make the environment more sustainable, decreasing the concentration of inert products in foods using insecticides rationally. However, many challenges must still be overcome, with that biotechnology has as a fundamental role, seek new research, sustainable in modern agriculture.

Importance of Biotechnology in the Control of Insects-prague

Biotechnology is defined as a set of techniques for manipulation of living beings or part thereof for economic purposes. This broad concept includes techniques that are used on a large scale in agriculture since the early 20th century, such as tissue culture, the biological fixation of nitrogen and organic pest control. But the concept includes also modern techniques of direct modification of the DNA of a plant or a living organism, in order to change precisely the characteristics of that organism.

Agricultural biotechnology varieties are used as a tool for agricultural research characterized by gene transfer of agronomic interest (and, consequently, of desired characteristics) between a donor agency (which may be a plant, a bacterium, fungus, etc.) and plants, safely.

Studies dating this year in India and China show that the Bt cotton production increased 10% to 50%, respectively, and the use of insecticides has reduced in both countries by 50%. In India, the producers have increased the income up to $ 250 or more per hectare, the farmer's income increasing national $ 840 million to $ 1.7 billion last year. Chinese farmers saw similar gains with increasing yields on average $ 220 per hectare, or more than $ 800 million nationwide.

It is important to recognize the need for scientific research that point on the General mode of action of Bt toxins and also that their toxicity is influenced by several factors. However, it is known that, in General, the toxicity of the Bt toxin on target organisms depends on factors such as certain pH, proteases, and the receivers. On the other hand, and more specifically, the extrinsic factors can also influence and co-factors in specific efficacy of Bt toxin on target organisms resistant and/or can also have an impact on the selectivity and toxicity to non-target organisms.

Therefore, the emergence of modern biotechnology marks the beginning of a new stage for agriculture and reserves a starring role to molecular genetics. The advances in the field of plant genetics have the effect of reducing the excessive reliance on agriculture mechanical and chemical innovations, which were the pillars of the green revolution. In addition to increased productivity, modern biotechnology can contribute to the reduction of production costs, better quality foods and for the development of less aggressive to the environment.

Bacillus Thuringiensis (Bt)

The Bacillus thuringiensis, was discovered in 1901, by the occurrence of an epidemic of mortality of larvae of the silkworm in Japan. Researchers found that it was caused by a previously unknown bacterium.

The bacterium entomopathogenic (Bt) stands out on the world stage since 1938, when the first product formulated with this pathogen was released in France.

In 1911, in Germany, the Berliner managed to isolate and characterize this bacterium, baptizing it Bacillus (by its cylindrical shape) thuringiensis (named after the German region of Turíngea). In 1938, France formulations containing right-handed bacteria colonies were sold as insecticides and, in 1954, its mode of action was discovered and its use today.

In some studies, this bacterium was considered inefficient in controlling S. frugiperda. However, with the advances provided by new laboratory techniques and greater interest of researchers' positive results were obtained.

The Bt a soil bacteria present in various continents, Gram-positive, aerobic and family Bacillaceae, when environmental conditions become adverse can sporulate to survive these conditions in. Are found in every terrestrial environments and also in dead insects, plants and debris. The number of cells obtained from Bt varied between 102 and 104 colony-forming units (CFU) per gram of soil, while in plants this number varies between 0 and ufc 100 cm^{-2}.

Produces sporangia containing a endospore and crystalline inclusions of proteins that are responsible for their action entomopathogenic, among which stands out the protein CRY. This crystal is composed of a protein polypeptide called endotoxin. When larval forms of insects feed on such proteins, initiates a series of reactions that culminate with the death of the same.

Biotechnology vs. Insecticide

The insects have been one of the major causes of damage to food production and, in world terms, the losses caused by pests and diseases are quite high. The same causing losses of the order of 38%. Withdrawals in Brazil indicate that pests can be liable for loss of 2.2 billion dollars for the main Brazilian crops.

Control of harmful insects is done, most of the time, by agrochemicals and, on a much smaller scale, by the employment of biological insecticides. The indiscriminate use of pesticides in combating the causal agents causes, despite its efficiency, environmental problems severe, human health, reduces number of natural enemies, and provides an accelerated selection of resistant insects. In contrast, biopesticides, Bt based, used for over a century, retainers of features less impactful on the environment and less harmful to humans ever occupied a prominent place on the market for the sale of pesticides.

From the Decade of 80, because of genetic advance, it became possible to develop a new pest control strategy, which consists of the genetically modified plants resistant to insects in and with effectiveness similar to conventional insecticides.

The first experiments with genetically modified (GM) plants were made in 1986, in the United States and in France. The first variety marketed a vegetable species produced by genetic engineering was the "FlavrSavr Tomato ", developed by the American company Calgene and marketed from 1994.

Between 1987 and 2000 there were more than 11,000 field trials in 45 countries and cultures more frequently tested were corn, tomatoes, soybeans, canola, potatoes and cotton, and the genetic features introduced were herbicide tolerance, product quality, virus-resistance and resistance to insects.

These days, according to the annual report of 2011 on the use of transgenic crops, the non-profit organization International Service for the acquisition of AgriBiotech Applications (ISAAA) observed an increase of 94 times in planted area of 1.7 million hectares in 1996 to 160 million hectares in 2011, allowing biotech crops become more agricultural technology adopted in the history of modern agriculture.

The endless search for alternative methods of insect control-Prague has been held strongly by several research groups worldwide, due to the need of a more sustainable agriculture and more committed to environmental preservation.

In this way, farmers have adopted this technology Bt targeting an increasing effective production to sustainable agriculture. The benefits of this technology are: reduction of environmental effects on toxins, safety in use, efficiency, conservation of natural enemies and reduction of fungal diseases.

The first advantage is the production of protein Cry, by plants-Bt, which is not affected by environmental factors such as atmospheric fallout, light incidence, and high temperatures. In addition, the homogeneity of the protein, in plant tissues, allows a more efficient use of insecticide effect than the application (spraying) of biopesticides, Bt, based on plants. The second advantage is the possibility of a higher level of security in relation to insecticide formulated because the proteins and does not accumulate in fatty tissues, are not toxic to humans and pets. Tied to these characteristics, the protein Cry, has no activity by contact, being necessary, the ingestion of the toxin by the insect, to have the effect of insecticide. The third advantage is the Heliothisvirescenscontrol significant and Pectinophoragossypiella, for example, in Bt cotton culture, between 95 and 99% efficiency. The fourth advantage is the preservation of natural enemies; therefore, secondary pests can become a problem if the population of beneficial insects is reduced by the use of chemical insecticides of low selectivity. The fifth benefit, no less important, is the reduction of fungal diseases. The lesions caused by insects, in the organs of plants, fungi infection, create opportunities mainly in the genus Fusarium e Aspergillusin. The primary importance of these fungi is the presence of micotoxins, particularly fumosinsandaflatoxins produced by them. The fumosins can be fatal to horses and pigs. Aflatoxinis extremely toxic to animals and humans. The dramatic reduction of insect attack, leads to reduction of insect attack, and consequently, decreases the production of micotoxins.

Mode of Action of Protein Cry

Currently the insect resistant transgenic plants expressing genes, inductors, an insecticidal protein called Cry, derived from the bacterium Bacillus thuringiensis (Bt). The mechanisms by which proteins exert their effect are Cryelucidated by pore formation model discussed below:

The mode of action of Cry proteins, produced by the plant, it is accomplished, orally, by susceptible insect. The process begins by solubilization of crystals in alkaline pH around 9.5, in the gut of insects, releasing protoxin of 130 kDa to Cry1 and Cry2 to 79kDa. After this breakdown, the protoxinare activated by digestive enzymes, forming toxic fragments of 60-65 kDa. These monomers bind to receptors specific primary, located in the apical membrane of the microvillus membranes of the columnar cells of the intestine of the larva. It is in this step that the affinity between the toxin and the receiver, for example, Cry1ae protein, lepdopteros is recognized as an important factor in determining the spectrum insecticidal Cry proteins. Later, the monomers bind to secondary receivers, which are proteins ancoradorasglicosil-phosphatidyl-inisitol (API), as phosphates and alkaline, to the lepidopteron Heliothisvirences. After this binding, the now oligomerose inserts into the membrane, where there are receptors for API, and leads to the formation of pores in the cell membrane of the intestinal epithelium and therefore destruction of microvilli membranes, hypertrophy of epithelial cells, vacuolization of cytoplasm, cell lyses and intestinal paralysis/death of the insect.

Safety of the use of Bt Plants

In relation to the safety of the use of Bt corn plants as an example, several tests are conducted to certify the safety of its use in the environment and in food and feed. Initially, the protein Cry is tested in animal models, such as rats and mice, for the verification of the toxic potential. One of these tests, called acute toxicity consists in forced ingestion of a pure animal protein by solution and on the observation of effects of this. The product only goes to the next steps of assessment if no effect is observed and diagnosed.

We can cite as an example the test performed with Safety, this protein in maize Herculex. This protein was tested on mice to the level of 576 mg per kilogram of live weight and no side effect was observed. To be exhibited at similar level, a person weighing 70 kg would take almost 5 tons of raw corn grains. This without taking into consideration the aspect that the human digestive tract, not to have alkaline pH, would not be able to downgrade this protein crystal.

Other reviews include the potential to cause allergies as well as the corn grain consumption by other animals such as chicken and fish, and what is called substantial equivalence, which is comparing the nutritional profile of the genetically modified maize with conventional maize. The corn will only be released commercially and, therefore, will go to the market when, in these analyses, the nutritional content between the conventional and transgenic corn were exactly the same, except, of course, the presence of protein inserted.

In the analysis of environmental safety, non-target organisms, how insects from another order, class or species, natural enemies and beneficial insects like bees, for example, are exposed to proteins inserted or the pollen grains that express and are evaluated its effects.

If all tests present results within expected ranges and be proven that there is no risk of harm or damage to health and the environment, these damages are compiled and submitted to the competent authorities of the country where you intend to market the product for analysis and approval of use and consumption. In Brazil these analyses are made by the national technical Commission on Biosafety (CTNBio) and the approval of a product in one country does not guarantee that the same is approved in another. For example, the event MON810 (Yield Gard) was approved in 1996 in the United States, in 1998 in Argentina and only in 2007 in Brazil.

For the use of Bt corn, just the producer, in addition to using the seeds of biotech corn, fulfill two rules: The coexistence, required by law, and the rule of Insect Resistance management (MRI), recommended by (CTNBio).

The coexistence rule requires the use of a 100 m isolating surround of transgenic maize plantations of corn to retain without transgenic contamination. Alternatively, you can use a surround of 20 m, provided they are sown maize transgenic not 10 ranks (equal-sized and transgenic maize cycle) isolating the area of transgenic maize.

The CTN Bio recommendation for Insect Resistance management is the use of the area of refuge. This recommendation is the result of consensus that the cultivation of Bt corn in large areas will result in the selection of biotypes of target pests resistant to Bt toxins. Obviously, the monitoring of the infestation of plants is also important because, depending on the used hybrid and intensity of infestation, the producer may need to adopt additional control measures.

The biggest concern with the use of Bt corn is on transgenic crops and coexistence of transgenic crops do not. Coexistence is the set of agricultural practices allowing farmers grain production from conventional transgenic and organic crops, according to standards of purity and to meet legal requirements for labeling. The adoption of the rules of coexistence is essential to preserve the freedom of choice of producers and consumers. Coexistence is also a topic particularly relevant when there is market incentive for the provision of non-transgenic maize. Evidence of their practical viability is the coexistence of a considerable number of different varieties of open pollination still in use.

Showed that companies in possession of this technology must guide growers on the rules of coexistence. The producer also held technical information, stick to them properly and conscious.

Information on packages of seed of Bt corn, there is a contract in which the producer, to open it, assumes the responsibility of following the rules of coexistence and the resistance management. Therefore, it is incumbent upon the producer responsibility of use of these rules. It is important to remember that the incorrect use of technology can take it to ineffectiveness in little time. If the producer is interested in paying more for Bt corn seed, is because he believes in the benefits that this technology is bringing to your production system. Therefore, it must be motivated to use this technology in a responsible way (using the area of refuge), to take ownership of this benefit for much longer.

In relation to Bt cotton to China is the leader in this technology. In 2006 6.3 million farmers, or more than 60% of the number of farmers who have sown transgenic in the world.

China is one of the only exceptions in the world to require shelters, although this may be changing. Even though the refuges were a way to reduce the accumulation of resistance to Bt toxin, the large number of small properties makes this strategy is difficult to apply.

The use of refuge in a developing country such as China becomes a challenging activity. Studies on policies of refuge on a large scale, in extensive agricultural systems of the United States of America, show that monitoring and implementation costs are negligible. Although this practice is reasonable in extensive production systems and with a small number of farms, they may not be suitable in developing countries. In developing countries like China, the agriculture sector is fragmented into millions of smallholdings, where each family has a diverse set of cultures.

As a result, it is likely that the implementation of the strategy of refuge to the style (IE, all farmers planting Bt cotton are forced to grow cotton non-Bt with refuge) would require a large implementation effort, making these types of strategies of refuge becomes unviable unless farmers received individual incentives to implement refuges based on self-interest. This is unlikely, since the build-up of resistance to Bt technology is a collective evil (compared to the more common public as well) that is unlikely to be accounted for by individuals in.

The area of refuge is the sowing of 10% of the area planted with Bt corn hybrids using Btnot equal size and cycle, preferably their isogenic hosts. The area of refuge should not be more than 800 m away from the transgenic plants. This is the average distance by dispersion of adults of LCM in the field in.

All these recommendations are in order to synchronize the intersections of potential adults surviving in area of Bt corn with which emergency in the area of refuge. The structured refuge must be drawn according to the acreage with the Bt corn to plots dimensions above 800 m in the shortest side (or Ray), cultured with Bt corn refuge will be needed in their tracks internal plots. Yet, according to the recommendation, in the area of refuge CTN Bio is allowed the use of other methods of control, provided that they are not used Btbased bio-insecticide.

BIOPESTICIDES

Biopesticides, a contraction of 'biological pesticides', include several types of pest management intervention: through predatory, parasitic, or chemical relationships. The term has been associated historically with [biological control] – and by implication – the manipulation of living organisms. Regulatory positions can be influenced by public perceptions, thus:

- In the EU, biopesticides have been defined as "a form of pesticide based on micro-organisms or natural products".

- The US EPA states that they "include naturally occurring substances that control pests (biochemical pesticides), microorganisms that control pests (microbial pesticides), and pesticidal substances produced by plants containing added genetic material (plant-incorporated protectants) or PIPs".

They are obtained from organisms including plants, bacteria and other microbes, fungi, nematodes, etc. They are often important components of integrated pest management (IPM) programmes, and have received much practical attention as substitutes to synthetic chemical plant protection products (PPPs).

Types

Biopesticides can be classified into these classes:

- Microbial pesticides which consist of bacteria, entomopathogenic fungi or viruses (and sometimes includes the metabolites that bacteria or fungi produce). Entomopathogenic nematodes are also often classed as microbial pesticides, even though they are multi-cellular.
- Bio-derived chemicals. Four groups are in commercial use: pyrethrum, rotenone, neem oil, and various essential oils are naturally occurring substances that control (or monitor in the case of pheromones) pests and microbial diseases.
- Plant-incorporated protectants (PIPs) have genetic material from other species incorporated into their genetic material (i.e. GM crops). Their use is controversial, especially in many European countries.
- RNAi pesticides, some of which are topical and some of which are absorbed by the crop.

Biopesticides have usually no known function in photosynthesis, growth or other basic aspects of plant physiology. Instead, they are active against biological pests. Many chemical compounds have been identified that are produced by plants to protect them from pests so they are called antifeedants. These materials are biodegradable and renewable alternatives, which can be economical for practical use. Organic farming systems embraces this approach to pest control.

RNA

RNA interference is under study for possible use as a spray-on insecticide by multiple companies, including Monsanto, Syngenta, and Bayer. Such sprays do not modify the genome of the target plant. The RNA could be modified to maintain its effectiveness as target species evolve tolerance to the original. RNA is a relatively fragile molecule that generally degrades within days or weeks of application. Monsanto estimated costs to be on the order of $5/acre.

RNAi has been used to target weeds that tolerate Monsanto's Roundup herbicide. RNAi mixed with a silicone surfactant that let the RNA molecules enter air-exchange holes in the plant's surface that disrupted the gene for tolerance, affecting it long enough to let the herbicide work. This strategy would allow the continued use of glyphosate-based herbicides, but would not per se assist a herbicide rotation strategy that relied on alternating Roundup with others.

They can be made with enough precision to kill some insect species, while not harming others. Monsanto is also developing an RNA spray to kill potato beetles One challenge is to make it linger on the plant for a week, even if it's raining. The Potato beetle has become resistant to more than 60 conventional insecticides.

Monsanto lobbied the U.S. EPA to exempt RNAi pesticide products from any specific regulations (beyond those that apply to all pesticides) and be exempted from rodent toxicity, allergenicity and residual environmental testing. In 2014 an EPA advisory group found little evidence of a risk to people from eating RNA.

However, in 2012, the Australian Safe Food Foundation posited that the RNA trigger designed to change the starch content of wheat might interfere with the gene for a human liver enzyme. Supporters countered that RNA does not appear to make it past human saliva or stomach acids. The US National Honey Bee Advisory Board told EPA that using RNAi would put natural systems at "the epitome of risk". The beekeepers cautioned that pollinators could be hurt by unintended effects and that the genomes of many insects are still unknown. Other unassessed risks include ecological (given the need for sustained presence for herbicide and other applications) and the possible for RNA drift across species boundaries.

Monsanto has invested in multiple companies for their RNA expertise, including Beeologics (for RNA that kills a parasitic mite that infests hives and for manufacturing technology) and Preceres (nanoparticle lipidoid coatings) and licensed technology from Alnylam and Tekmira. In 2012 Syngenta acquired Devgen, a European RNA partner. Startup Forrest Innovations is investigating RNAi as a solution to citrus greening disease that in 2014 caused 22 percent of oranges in Florida to fall off the trees.

Bacillus thuringiensis, a bacteria capable of causing disease of Lepidoptera, Coleoptera and Diptera, is a well-known insecticide example. The toxin from B. thuringiensis (Bt toxin) has been incorporated directly into plants through the use of genetic engineering. The use of Bt Toxin is particularly controversial. Its manufacturers claim it has little effect on other organisms, and is more environmentally friendly than synthetic pesticides.

Other microbial control agents include products based on:

- Entomopathogenic fungi (e.g. Beauveria bassiana, Isaria fumosorosea, Lecanicillium and Metarhizium spp).

- Plant disease control agents: include Trichoderma spp. and Ampelomyces quisqualis (a hyper-parasite of grape powdery mildew); Bacillus subtilis is also used to control plant pathogens.
- Beneficial nematodes attacking insect (e.g. Steinernema feltiae) or slug (e.g. Phasmarhabditis hermaphrodita) pests.
- Entomopathogenic viruses (e.g. Cydia pomonella granulovirus).
- Weeds and rodents have also been controlled with microbial agents.

Various naturally occurring materials, including fungal and plant extracts, have been described as biopesticides. Products in this category include:

- Insect pheromones and other semiochemicals.
- Fermentation products such as Spinosad (a macro-cyclic lactone).
- Chitosan: A plant in the presence of this product will naturally induce systemic resistance (ISR) to allow the plant to defend itself against disease, pathogens and pests.
- Biopesticides may include natural plant-derived products, which include alkaloids, terpenoids, phenolics and other secondary chemicals. Certain vegetable oils such as canola oil are known to have pesticidal properties. Products based on extracts of plants such as garlic have now been registered in the EU and elsewhere.

Applications

Biopesticides are biological or biologically-derived agents, that are usually applied in a manner similar to chemical pesticides, but achieve pest management in an environmentally friendly way. With all pest management products, but especially microbial agents, effective control requires appropriate formulation and application.

Biopesticides for use against crop diseases have already established themselves on a variety of crops. For example, biopesticides already play an important role in controlling downy mildew diseases. Their benefits include: A 0-Day Pre-Harvest Interval, the ability to use under moderate to severe disease pressure, and the ability to use as a tank mix or in a rotational program with other registered fungicides. Because some market studies estimate that as much as 20% of global fungicide sales are directed at downy mildew diseases, the integration of biofungicides into grape production has substantial benefits in terms of extending the useful life of other fungicides, especially those in the reduced-risk category.

A major growth area for biopesticides is in the area of seed treatments and soil amendments. Fungicidal and biofungicidal seed treatments are used to control soil borne fungal pathogens that cause seed rots, damping-off, root rot and seedling blights. They can also be used to control internal seed–borne fungal pathogens as well as fungal pathogens that are on the surface of the seed. Many biofungicidal products also show capacities to stimulate plant host defence and other physiological processes that can make treated crops more resistant to a variety of biotic and abiotic stresses.

Disadvantages

- High specificity: Which may require an exact identification of the pest/pathogen and the use of multiple products to be used; although this can also be an advantage in that the biopesticide is less likely to harm species other than the target?
- Often slow speed of action (thus making them unsuitable if a pest outbreak is an immediate threat to a crop).
- Often variable efficacy due to the influences of various biotic and abiotic factors (since some biopesticides are living organisms, which bring about pest/pathogen control by multiplying within or nearby the target pest/pathogen).
- Living organisms evolve and increase their resistance to biological, chemical, physical or any other form of control. If the target population is not exterminated or rendered incapable of reproduction, the surviving population can acquire a tolerance of whatever pressures are brought to bear, resulting in an evolutionary arms race.
- Unintended consequences: Studies have found broad spectrum biopesticides have lethal and nonlethal risks for non-target native pollinators such as Melipona quadrifasciata in Brazil.

BIOFUNGICIDES

The term biofungicide can have several different meanings, but it is most frequently used to refer to fungicides that contain a microorganism (usually a bacterium or fungus) as the active ingredient.

These microbially-based biofungicides are the focus of this article. Biofungicides can control many different kinds of fungi and water molds, although each separate active ingredient controls only certain pathogens. Some also control bacterial diseases. Virtually all of the organisms used in biofungicides on the market today occur naturally in soil or on plant surfaces, and most are approved for use in organic production.

Advances in fermentation technology have allowed mass production of highly specialized microbes that previously could only be grown in small batches on highly specific substrates, such as on roots infected with pathogens. Consumer demand for organically certifiable pesticides and increased regulatory pressure on older synthetic pesticides, especially in Europe, has fostered increased commercial interest in the production of living organisms that can suppress or kill pathogens. For these reasons and also because as natural products biofungicides generally have few negative impacts on health and the environment, the number available will likely continue to increase.

Powdery mildew in roses can be managed with foliar applications of the biofungicide Bacillus subtilis.

Since microbial biofungicides contain living organisms, their modes of action differ from those of synthetic fungicides. Some of these mechanisms include:

- Competition: The biocontrol agent is more effective than the pathogen at gathering critical nutrients or space and, therefore, must be in place before disease onset.
- Antibiosis: The biocontrol agent produces a chemical compound of some type (antibiotic or toxin) that acts against the pathogen.
- Predation or parasitism: The biocontrol agent directly attacks the pathogen.
- Induction of host plant resistance: The biocontrol agent triggers a defensive response in the host plant that limits the ability of the pathogen to invade the plant.

Most biofungicides use one or more of the above mechanisms to target only one or a few specific pests. As such, applicators should both read the label and diagnose the problem carefully to insure that the product will be effective. Biofungicides work best when applied preventively. Application after a plant is already infected has little chance of significantly altering the course of the disease for that plant, although it may decrease the ability of the pathogen to move from that plant to other plants, especially if the pathogen has to move through the soil to do so. Thus, an application of biofungicide is not likely to cure an infected plant; but it may protect other nearby plants in the field.

Black spot on rose can also be managed with foliar applications of Bacillus subtilis.

Biofungicides containing Streptomyces or Trichoderma can be used to prevent infection of plants with damping off or root or seed rot pathogens such as Pythium.

Although independent testing by university researchers and others has verified some manufacturer claims made for these products, efficacy data for many other products against pathogens on ornamentals is unavailable. For instance, as of early 2014, independent testing has not demonstrated adequate field efficacy of any biofungicide for landscape or agricultural use against Armillaria root rot (otherwise known as oak root rot). However, a number of biologicals have been found effective for control of Pythium, Phytophthora, Verticillium, and other pathogens on a variety of plant hosts. With these types of products, eradication of the pathogen is not the goal and is probably never achieved. Instead, biofungicides rely on a core tenet of the IPM philosophy: Keeping pest levels below damaging thresholds and using biofungicides (when necessary) in combination with cultural practices that promote healthy plant growth.

Biofungicides cannot take the place of proper cultural care. They are a valuable tool for keeping a strong plant healthy, but they cannot forestall the inevitable. If your client's Japanese maples are routinely drowned, allowed to wilt, and then drowned again, adding a biofungicide will not prevent them from contracting Phytophthora if it is present in the soil.

If biofungicides are a useful and environmentally friendly tool in the landscape, why aren't they more widely used? One reason is that these fungicides rely on living organisms for efficacy, so they must be stored appropriately in order to retain their fungicidal properties over time. However, a more commonly cited reason is that the personal protective equipment needed to apply them is more involved than for some other compounds. Routine exposure to the proteins found in the spray mists of some biofungicides can result in the development of allergic reactions. To keep commercial applicators safe, they must wear NIOSH approved respirators when mixing, loading, or applying biofungicides in agricultural or landscape settings. This may not be immediately obvious when reading the labels, as a quick scan often only shows the following required personal protective equipment (PPE):

- Long sleeved shirt and long pants.
- Shoes plus socks.
- Waterproof gloves.

The respirator requirement is only evident when reading the text following the list. Biofungicides are safe to use as long as mixer/loaders and applicators have and use a respirator as part of their PPE. However, a NIOSH approved respirator requires proper training and fitting in order to be effective.

When used properly and with forethought, biofungicides can be an important part of an IPM program to prevent or mitigate problems with plant pathogens in the landscape. However, nothing can ultimately take the place of proper plant selection and care.

6

Bioinformatics in Agriculture

Bioinformatics is the discipline of science that deals with the study of information technology to manage biological data for decoding plant genomes. It has aided in genome sequencing for studying protein structures, increasing plant yield, improving nutritional quality, resisting insects and pests, etc. This chapter closely examines the role of bioinformatics in agriculture for an extensive understanding of the subject.

The key role of bioinformatics is acquiring striking importance in the era of outstanding advances in omics technologies for its fundamental support in describing the multifaceted aspects of biological functionalities. The manifold omics efforts flourishing worldwide are also contributing fundamental novelties in many aspects of agricultural sciences and, as a consequence, bioinformatics is acquiring a crucial role also in these research fields. Indeed, the transformation of natural environment for improvement of goods from animal, plants, and microbial worlds for human nutrition and health requires the comprehension of the molecular mechanisms influencing the structure and the function of the individuals, the populations, and the communities. The expanding knowledge about the molecules and the mechanisms associated with specific phenotypic traits and specific responses to biotic or abiotic stresses, complemented with the predictive power of bioinformatics, has an impact on agriculture practices and favors innovative methods in diagnostics, monitoring, and traceability, improving human benefits at lower costs, thus supporting sustainability.

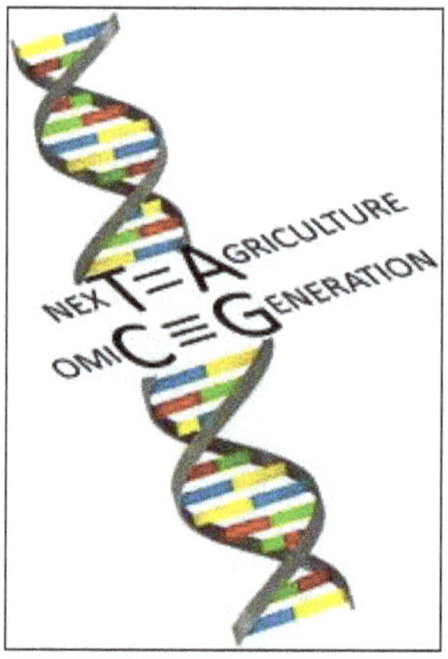

Bioinformatics emerged from the initial requirement of suitable informatics for biological data organization, management, and distribution, but soon it revealed also fundamental in providing tools for data analysis, interpretation, and modeling. Moreover, thanks to bioinformatics, it was possible to analyze and understand structure and function not only of single bio-molecules but also of larger molecular collections, derived from the so-called omics experimental approaches. These efforts permit to depict different aspects (genomics, transcriptomics, proteomics, metabolomics, etc.) of the biomolecular organization of complex biological systems, from cells to ecosystems. The fast spreading of omics techniques, with its growing power and more accessible costs, drastically increased the amount of molecular data collections from different levels of organization of an organism or an environmental sample. This favored a holistic view on systems organization and functionality, further challenging bioinformatics with data size and the need of integrative efforts. The recent introduction of Next-Generation Sequencing (NGS) technologies. Further revolutionized the sequencing of nucleic acids contributing to a new era in omics approaches. Indeed, on one hand these technologies introduced an incredible efficiency in terms of experimental execution time and a deeper resolution. On the other hand, they stimulated an unexpected interest from scientists due to the higher affordability in terms of experimental procedures and economical requirements. We here introduce the novelties that the advent of NGS technologies contributed in agriculture, overviewing the main bioinformatics strategies and challenges, as well as perspectives in the field.

Single and Multi Species Genomics for Agriculture

Genomics, transcriptomics, proteomics, and metabolomics may contribute to the comprehension of the organization and the functionality of biological systems, with the possibility to also trace molecular variability during development, in different conditions, such as physiological, pathological, or influenced by environmental changes.

Samples for omics studies can derive from one or several individuals of a species (a population), or from multiple species (a community). The differences among these approaches consist mainly in the objectives of the specific studies.

In single individual approaches, the organization and the functionality of specific cells, tissues, or organs (e.g., roots, fruit, rumen) are investigated, mainly to identify factors influencing emergent properties, like the quality and the shape. This also paves the way to the characterization of even more complex traits (e.g., yield, resistance to stresses, diseases) or processes (e.g., fruit ripening, growth efficiency, senescence).

The description of the molecular components in a population of the same species aims to the understanding of evolutionary processes influencing genetic variability. This can also widely contribute to dissect complex quantitative traits by identifying novel and superior alleles or to assess the impact of genetic variation on patterns of gene expression and on phenotypic plasticity in response to environmental changes.

The study of the collective genetic pool deriving from communities is termed "metagenomics", a term dating back to 1998. The community can derive from environmental samples, such as soil, seawater, or other, but also part of individuals, such as gut or roots. Metagenomics usually aims to describe the prokaryotic component of the community, but may be also useful to trace the different eukaryotes existing in a specific environment.

Nucleic acid sequencing always contributed the majority of the data in all the approaches here summarized. This is why the recent introduction of NGS technologies impressively impacted the productivity and the advancement in these research fields.

Impact of NGS in Agriculture

The multifaceted scientific topics in agriculture sciences may be consistently supported by NGS omics for single individuals, populations, or communities.

The sequencing of whole genomes from several species permits to define their organization and provides the starting point for understanding their functionality, therefore favoring human agriculture practice. Efforts addressed to the achievement of an appropriate knowledge of associated molecular information, such as the one arising from transcriptome and proteome sequencing, are also essential to better depict the gene content of a genome and its main functionalities. These efforts indeed led to major advancements in all biological sciences and in agriculture as well. Moreover, the elucidation of the complexity of genes and their networking is also fundamental for being eventually translated into breeding practice for crops or livestock, contributing to their health, resistance, and productivity. Indeed, the contribution of genomics to agriculture spans the identification and the manipulation of genes linked to specific phenotypic traits as well as genomics breeding by marker-assisted selection of variants. The so-called "agricultural genomics" (or agri-genomics), indeed, aims to find innovative solutions through the study of crops or livestock genomes, achieving information for protection and sustainable productivity for food industry, but also for alternative aspects like energy production or design of pharmaceuticals.

Plant, soil, and livestock microbiome also play a key role in agriculture since it determines plant fitness, soil biogeochemical properties, and affects both yield and quality traits. However, little knowledge is available for microbes and the communities in which they are included. As an example, it is acknowledged that soil is one of the biggest carbon reservoirs on earth, and prokaryotes constitute an important amount of the soil biomass. However, culture-independent studies in the last three decades showed that, although sequencing strategies are fast evolving, the great majority of bacterial species is still unknown. Therefore most of the methods used for profiling microbial communities and describe their main functional features are now adopting whole DNA extraction and the use of NGS on the entire sample, with the objective of sequencing and characterizing DNA fragments of all the species included, i.e., the metagenome. The application of metagenomics in agriculture also proved to be appropriate for depicting the complex patterns of interactions occurring among micro-organisms in soil and in plant rhizosphere, as well as in specific tissues or organs. Metagenomics recently revealed to be useful to trace the shift in taxonomic composition and functional redundancy of microbial communities in rhizosphere and in soil in connection to environmental changes associated to fertilization and agricultural management. Metagenomics studies can also help deciphering the role of soil bacteria in plant nutrition or in the cycle of the elements. Further applications can lead to the discovery of new genes, bio-products, plant growth promoting micro-organisms consortia, useful for understanding relevant aspects such as response to stresses or dysbiosis.

The Revolution in the Omics Technologies and the Impact on Bioinformatics

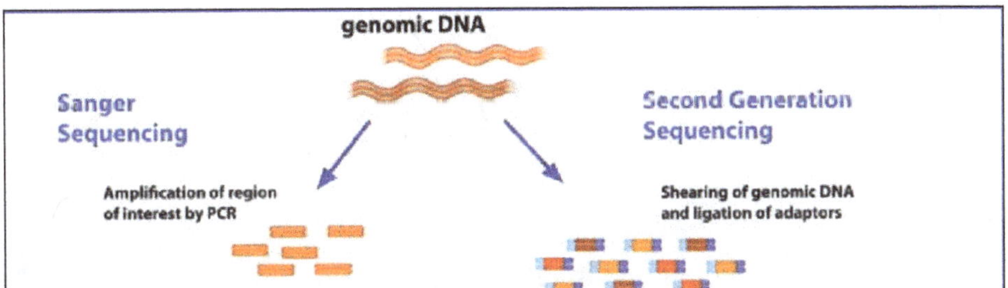

The release of the completely sequenced genomes for some of the major plants (green), animals (red), and fishes (black) ofinterest in agriculture. The dashed line indicates the start of Next-Generation Sequencing (NGS) era.

The introduction of omics approaches strongly impacted bioinformatics in data collection, organization, integration, and in the implementation of suitable data mining tools. The support of efficient bioinformatics favored the introduction of the so-called high-throughput technologies, paving the way to the flourishing of genome sequencing efforts of key model species, such as Homo sapiens and Arabidopsis thaliana. The same technologies were then exploited to further push forward the genome sequencing of other model and non-model species, many of which of agriculture interest. These efforts were also preceded or accompanied by transcriptome sequencing efforts using different technologies, in support of gene prediction, but also for depicting transcriptional processes and define cell functionality in physiological, pathological, or stress conditions. These approaches also required the design of appropriate resources to distribute the data and/or dedicated collections of processed results to all the interested scientific community, enhancing the need for suitable pipelines for moving from raw to value added information and integrative data mining.

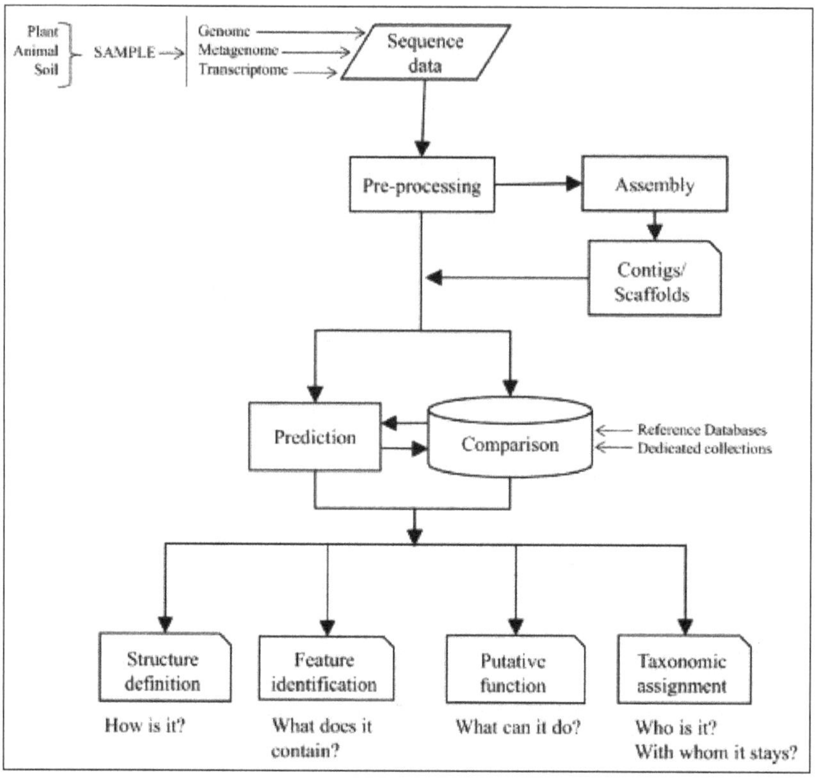

General description of a standard workflow in omics data analyses. Starting from the biological sample of interest and sequencing by different approaches data can be processed by assembly into contigs/scaffolds or directly undergo prediction analyses and/ or comparison with reference databases. The analyses support different kinds of investigations such as those required for (i) structure definition, (ii) feature identification (e.g., genes, regulatory and repeat regions), (iii) putative functionality, and (iv) taxonomic assignment.

NGS strikingly contributed to expand the number of genomes currently completely sequenced, as well as to the establishment of novel ambitious efforts, for instance, those focused on multi-genome sequencing or those aiming to define global metagenomes from different environmental samples to define reference collections. These technologies are also exploited for the production of alternative, related collections, namely from transcriptomics, epigenomics, and metagenomics projects. The unexpected amount of raw data the new technologies are providing requires also dedicated storage for centralized data maintenance, currently solved by the SRA system. Worthy to note is the size reached by the SRA archive in a short time span when compared to the entire nucleotide collection currently available. NGS data size represents a big challenge for bioinformatics. Indeed, main computational tasks are today focused on the optimization and adaptation of typical methodologies in bioinformatics to the magnitude of NGS collections.

Data are usually delivered by sequencing centers in the form of raw, fragmented sequences, to be pre-processed, i.e., cleaned, from additional fragments due to the specific technology employed, such as vectors, adaptors, barcodes, or other contaminations. Structure definition from fragmented data usually requires an assembly step to reconstruct the most reliable original molecules, such as longer genomic sequences or transcripts. The assembly is based on sequence alignments driven by identical regions shared by the fragments. The assembly procedure may include already available reference sequences (guided approaches), as in the case of transcript assemblies based on a genome reference, or they are based on reference-free methodologies (de novo). This procedure is a key step for many different applications, indeed high-quality longer backbones are useful to properly proceed towards successive steps in data processing, which are mainly structure and functional assignments and predictions. In genomics, transcriptomics, population genetics, and metagenomics, these are widespread fundamental tasks solved by different computational methodologies, though based on similar strategies, mainly ab initio approaches or similarity-based ones. Ab initio-based algorithms exploit complex probabilistic models to detect expected features (genes, motives, propensities) as defined on the basis of training datasets that support the proper identification of specific features. Similarity-based methods, on the other hand, are the principal and more frequently used approaches in bioinformatics since they permit identifications, predictions, structure, and functional assignments. They rely on sequence or tridimensional similarities exploiting a typical concept in biology, which considers similarity in structure as a possible indication of similarity in role. Comparison with reference (nucleotide or amino acid general databases) or more specific collections, such as those from genomes, gene families, transcriptomes, or repeats, is fundamental to transfer information from already annotated molecules to newly defined ones. Beyond supporting structure and functional assignments by detection of common features, similarity searches also support identification of peculiarities and provide hints for evolutionary investigations.

The availability of an increasing number of reference genomes associated with a decreasing sequencing cost per base enabled also the analysis of genome variations based on single nucleotide polymorphisms (SNP) discovery. These studies intend to identify the variability between genomic sequences from individual genomes. By comparing the sequenced genomes, a catalog of mutations from individuals is obtained, usually defined as SNPs and/or insertion-deletion (INDELs), but also as larger rearrangements (e.g., copy number (CNV) and presence absence (PAV) variations, translocations). These features can be associated with specific phenotypes of interest. Currently, millions of polymorphisms have been discovered in plants, such as A. thaliana, rice, soybean, tomato, maize, and in animals. These resources are essential in breeding challenges for species of agriculture interest. Indeed, the possibility to exploit data from larger collections of individuals strongly increases the potential to identify more alleles useful for improved a sustainable productions, providing solutions for growing demand for better food, in a climate changing world.

As introduced, metagenomics approaches aim at the identification of the species in a sample and to the definition of their relationships. While bioinformatics approaches for what concerns pre-processing and assembly steps are similar, downstream analyses depend on the peculiarities of the implemented strategy. The taxonomic composition of the microbiome can be profiled using the marker-based approach, i.e., a PCR-amplification with universal primers of a taxon-specific gene, followed by the extensive sequencing of the amplicon by the preferred platform. Sequences derived from such studies are usually compared with dedicated databases representing high-quality full-length reference tags. Pipelines have been also implemented to aid non-experts users in a correct parsing of the metagenome-derived NGS data. Phylogenetic relationships, obtained by sequence similarity, can be used for ecological inference using dedicated pipelines. However, although widespread, the marker-based approach falls short in predicting the functionality and the activity of the microbial community. Indeed, the methodology suffers the typical PCR biases, such as (i) the misincorporation of nucleotides (which would lead to the overestimation of sequence diversity); (ii) the differential amplification of the same gene from different organisms (true for example in the case of 16S genes whose number of copies in the genome varies among taxa; and (iii) the formation of chimeric artifacts. Moreover, markers can have limits in taxonomical assignments, mainly because of lack of consistent genome information from all possible species, affecting the specificity of the identification of the components. This leads to the wide use of the operative taxonomic units (OTUs) for distinguishing all the different components in a sample, since they represent groups of highly similar sequences.

The "shotgun" approach is alternative to the marker-based one. It consists in the high-throughput sequencing of a pool of DNA fragments that may encompass various genomic loci from all taxa represented in the sample (prokaryotic, eukaryotic, and viral genomes). Unlike target-based approaches, the shotgun technique provides more details on the genomic structure of the community, offering a wider description of its potential functionality. Data from whole metagenome shotgun consist of short DNA reads that can be assembled to obtain coding sequences or genomic contigs. Coding sequences can be identified through the comparison with specific databases. The assembly should be carefully evaluated because most of the assemblers were developed for genome assembly and are not designed to deal with the heterogeneity of metagenomic datasets. In alternative, raw reads can be also used for direct assignment and annotation, though their short length may limit the exhaustivity of the results. Ultimately, other limits of the shotgun method are (i) the initial amount of extracted DNA for library production should be rather high (>10 ng); and (ii) in case of large and complex communities, or communities where one or few species dominate over the others, the coverage of the entire components may be limited. Indeed, the likelihood that the species poorly represented will be covered by sufficient reads that will also permit the assembly of representative contigs is rather low. Examples of main software dedicated to metagenome analyses. Ultimately, other limits of the shotgun method are (i) the initial amount of extracted DNA for library production is rather high (>10 ng); and (ii) in case of large and complex communities, or communities where one or few species dominate over the others, the likelihood that more than one read will cover a single gene is rather low; therefore, little information will be obtained about the species with low abundance, and assembly would probably result in short contigs (if any). As introduced, reference data collections are fundamental since the beginning of bioinformatics. Data sharing, by general reference or specialized databases, is precious to the majority of the bioinformatics strategies here presented. They not only support fundamental analyses for straightforward characterization of the investigated molecules, but are also essential to offer results to the whole scientific community. To this aim, the effort of setting up comprehensive collections, suitably representing the metadata derived by their mining and by the integration of different resources, demands major efforts in the NGS bioinformatics era. Indeed, NGS technologies attracted an unexpected interest from the scientific world for their accessibility and their resolution power, further challenging the stabilization of resources and data integration. As an example, the fast sequencing of complete genomes from different species, such as those from different genotypes or cultivars, faces the bottleneck caused by the need of suitable data analyses and curation. On the other hand, the fast production of collections from parallel efforts makes the update of novel release hard, though the presence of reference databases, favoring the flourishing of community specific collections, often misaligned with reference ones, affecting also the quality of the results.

Bioinformatics is the exclusive approach capable of exploiting and sharing the large amount of omics data the different technologies may provide. Suitable computational methods and appropriate resources are fundamental for detecting value added biological information providing novel insights into the organization of biological systems. The identification of structure and functional properties of the molecular data in a specific process allows the in-depth understanding of systems organization and behavior, supporting the design of reliable and representative models and paving the way to the comprehension of emergent properties that only holistic approaches can offer.

However, despite the introduction of highly processive experimental technologies and of innovative computational approaches in support of the molecular characterizations, only 10 % of the genome organization and associated functionalities have been today understood and an even lower percentage of metagenomics datasets have been confidently annotated. This confirms that though at quite 70 years from the discovery of the DNA structure and the beginning of bioinformatics, we are still at the very early stage of the genomics era, and surely quite far from achieving the ambitious goal of the in silico simulation of complex living organisms as well as ecosystem relationships. Indeed, these efforts still demand for extensive and suitable studies of genomes, transcriptomes, and metagenome data for proper links with sample organization and functionality, considering single species analyses and community approaches.

Despite these limits, the NGS bioinformatics era is revolutionizing the experimental design in molecular biology, strikingly contributing in increasing scientific knowledge while impacting relevant applications in many different aspects of agriculture. Data from disparate research fields, such as breeding, microbiology, and environmental sciences, are favoring a common exploitation and advances in molecular knowledge from massive efforts, with bioinformatics as driving methodology for its power and multifaceted capabilities. Organizing, detecting, integrating data information content, and data sharing are contributing to multidisciplinary interactions, expanding resources and spreading common methodologies. This revolutionize agriculture practice and production, offering knowledge and tools for improved product quality and ameliorated strategies of protection against environmental stress, diseases, and parasites. The different applications here overviewed, beyond providing relevant scientific knowledge based on their specificities, are also fundamental for translational approaches providing contributions with technological innovation, novel products, predictive and monitoring approaches, also supporting innovative applications for crop and livestock management.

The increase of omics-based studies needs education in the associated technologies and in bioinformatics for appropriate experimental design and analyses, and for properly conveying experimental and computational efforts towards an in-depth knowledge and appropriate modeling of the biological systems.

GENOMIC DATABASES FOR CROP IMPROVEMENT

The majority of DNA sequence and expressed gene sequence data generated today comes from the next- or second-generation sequencing (NGS/2GS) technologies. NGS technologies produce vast quantities of short data rather than Sanger sequencing at a relatively low cost and short time. Genomics is undergoing a revolution, driven by advances in DNA sequencing technology, and this data flood is having a major impact on approaches and strategies for crop improvement. NGS technologies have been applied for sequenced genomes of a number of cereal crop species including rice, Sorghum and maize. A quality sequence of rice that covers 95% of the 389 Mb genome has been produced. The Sorghum bicolor (L.) Moench genome has been assembled in size of 730-megabase, placing ~98% of genes in their chromosomal context. The draft nucleotide sequence of the 2.3-gigabase genome of maize has also been improved. One of the challenges encountered by researchers is to translate this abundance of data into improved crops in the field. There remains a gap between genome data production and next-generation crop improvement strategies, but this is being rapidly closed by far sighted companies and individuals with the ability to combine the ability to mine the genomic data with practical crop-improvement skills. Bioinformatics can be defined as the structuring of biological information to enable logical interrogation, and databases are a key part of the bioinformatics toolbox. Numerous databases have been developed for genomic data, on a range of platforms and to suite a variety of different purposes. These range from generic DNA sequence or molecular marker databases, to those hosting a variety of data for specific species.

Generic Databases

The largest of the DNA sequence repositories is the International Nucleotide Sequence Database Collaboration (INSDC), made up of the DNA Data Bank of Japan (DDBJ) at The National Institute of Genetics in Mishima, Japan, GenBank at the National Center of Biotechnology Information (NCBI) in Bethesda, USA, and the European Molecular Biology Laboratory (EMBL) Nucleotide Sequence Database, maintained at the European Bioinformatics Institute (EBI) in the UK. Daily data exchange between these groups ensures coordinated international coverage.

Since the introduction of advanced next-generation sequencing technology, the storage and interrogation of this data is becoming an expanding challenge. The ability to search the vast quantity of this data is made feasible by the development of custom databases such as TAGdb, but it is increasingly the assembled and annotated genome data which are applied for crop-improvement applications.

While it is valuable to maintain all public nucleic acid sequences in one location, the size of this resource limits the ability to visualize this data. Genome viewers, which place genomic data within the context of sequenced or partially sequenced genomes, provide more context-orientated data interrogation. There are two main generic web-based tools to view plant genomes: Ensemble and GBrowse. Both are widely used and it is not uncommon to find similar genome information hosted on both systems. A key development in genome databases was the establishment and adoption of a standard file format for genome data, and data in the current version, GFF3 can be visualized and searched using a wide range of tools from custom GBrowse databases to stand alone bioinformatics tools such as Biomatters Geneious.

There are several resources which collate genome data for multiple plant species. Gramene is an EnsEMBL-based genome viewer and database hosting information on a variety of crop species, but based around the rice, maize and Arabidopsis genomes. A similar resource is hosted by the EBI. PlantGDB is a resource for comparative plant genomics and hosts sequence data for >70,000 plant species with a focus on complete sequencing of reference species, Arabidopsis, rice, maize and Medicago truncatula. Plaza hosts pre-computed comparative genomics data sets for a range of species. Phytozome also hosts genome data for numerous plant species and provides several genomes using the GBrowse format. With 25 complete plant genomes, phytozome is one of the most comprehensive plant genome databases currently available. In addition, PlantsDB is a generic database hosting data for multiple plant species. This database is hosted by MIPS.

While genome and transcript sequence information makes up the bulk of genome data maintained within public databases, it is often the differences between individuals and varieties which are the most valuable for crop-improvement applications. A major focus of crop genetic research in recent decades has been the development of molecular genetic markers associated with important traits. Genetic markers can be assayed with a variety of techniques. Early molecular genetic markers technologies such as restriction fragment length polymorphisms have been replaced by more high throughput methods, including amplified fragment length polymorphisms (AFLPs), diversity array technologies (DArT) and simple sequence repeats (SSRs) also known as microsatellites. Another important and crop-improvement-oriented database is the maize database Panzea, which hosts data on genomic diversity in a large germplasm collection including genetic data, trait phenotypes, allele frequencies, phenotyping environments, genetic analysis tools and so on. The Panzea database The Panzea databases comprises the genotypic and phenotypic data and genetic marker information. This database design is based on the Genomic Diversity and Phenotype Data Model (GDPDM).

An expressed sequence tag (EST) represents a short sub-sequence of a cDNA sequence. EMBL or Gen Bank have sub-sections for EST sequences. The crop expressed sequence tag database, CR-EST, provides access to more than 200,000 sequences derived from 41 cDNA libraries of four species: Barley, wheat, pea and potato.

SSRs are short stretches of DNA sequence occurring as tandem repeats of mono-, di-, tri-, tetra-, penta- and hexa-nucleotides. They are highly polymorphic due to mutation affecting the number of repeat units. The value of SSRs is due to their genetic co-dominance, abundance, dispersal throughout the genome, multi-allelic variation and high reproducibility. The hypervariability of SSRs among related organisms makes them excellent markers for genotype identification, analysis of genetic diversity, phenotype mapping and marker assisted selection. SSRs demonstrate a high degree of transferability between species, as PCR primers designed to an SSR within one species frequently amplify a corresponding locus in related species, enabling comparative genetic and genomic analysis.

With the continued advances in DNA sequencing technologies, single-nucleotide polymorphisms (SNPs) have come to dominate high throughput molecular marker analysis. SNPs are the ultimate form of molecular genetic marker, as a nucleotide base is the smallest unit of inheritance, and a SNP represents a single-nucleotide difference between two individuals at a defined location. SNPs are direct markers as the sequence information provides the exact nature of the allelic variants. Furthermore, this sequence variation can have a major impact on how the organism develops and responds to the environment. SNPs represent the most frequent type of genetic polymorphism and may therefore provide a high density of markers near a locus of interest. SNPs at any particular site could in principle involve four different nucleotide variants, but in practice they are generally biallelic. This disadvantage, when compared with multiallelic markers such as SSRs, is compensated by the relative abundance of SNPs. The high density of SNPs makes them valuable for genome mapping, and in particular they allow the generation of ultra-high density genetic maps and haplotyping systems for genes or regions of interest, and map-based positional cloning. SNPs are used routinely in crop breeding programs, for genetic diversity analysis, cultivar identification, phylogenetic analysis, characterization of genetic resources and association with agronomic traits.

SSR Primer is a web-based tool that enables the real time discovery of SSRs within submitted DNA sequences, with the concomitant design of PCR primers for SSR amplification. Alternatively, users may browse an SSR Taxonomy Tree to identify pre-determined SSR amplification primers for species represented within the GenBank database.

The SNP discovery software auto SNP identifies SNPs and insertion/deletion (indel) polymorphisms from bulk sequence data using two measures of confidence; redundancy, defined as the number of times a polymorphism occurs at a locus in a sequence alignment; and co-segregation of SNPs to define a haplotype. AutoSNP software has recently been extended to database format, autoSNPdb, which permits complex queries and provides detailed genomic and functional information. Where the sequence trace files are available, the SNP discovery tool Poly Phred can make use of the base pair quality scores to further differentiate between true SNP polymorphisms and random sequence error. The recent developments in next-generation sequence data have led to the identification of large numbers of SNPs in a range of plant genomes and these approaches are likely to dominate SNP discovery in the coming years.

The increased throughput for the discovery and application of molecular genetic markers has led to the requirement for databases hosting the results of molecular marker analysis. These maybe integrated within other database systems such as Gramene, the Legume Information System (LIS), or Graingenes.

One of the principal uses of molecular genetic markers is the production of genetic maps and the mapping of heritable traits. While mapping data may be described as lists, graphical representations are more readily understood. The genetic map viewer CMap, developed by the GMOD consortium is valuable for the validation of traits that map to the same position in different populations and also for the linkage between crop genetic maps and sequenced model genomes, enabling the identification of candidate genes for genetically mapped traits. A recent addition, CMap3D, enables the comparison of a larger number of maps in 3D space.

The linking of genomic data with agronomic traits remains one of the greatest challenges in the application of genome data for crop improvement. Several databases have been developed to assist in this endeavor. The International Crop Information System (ICIS) is a database system that hosts integrated management information for crop improvement, including details on diverse germplasm and traits. One challenge in developing trait databases is the establishment of functional ontologies. The Plant Ontology is a controlled vocabulary (ontology) that describes plant anatomy and morphology and stages of growth and development for all plants and this database of ontologies is becoming the standard for comparative physiology and for linking genes with potential function.

Species Focused Databases

It would be impossible to detail all available plant genetic and genomic databases, however some of the main ones are listed below along with a brief description of their content.

GrainGenes is a genetic database for Triticeae, oats, and sugarcane GrainGenes. Comprehensive information includes genetic markers, map locations, alleles, key references and disease symptoms. The Triticeae Repeat Sequence Database (TREP) contains a collection of repetitive DNA sequences from different Triticeae species which can be used for the development of molecular markers.

While Brachypodium distachyon is not grown as a crop, this species has many qualities that make it a model for studies in temperate grasses and cereals, including a small genome (~ 300 Mbp), small physical stature, self-fertility, a short lifecycle, and simple growth requirements. The B. distachyon genome was sequenced in 2010 and the Brachypodium database which includes a GBrowse based genome viewer is available.

The maize genome was sequenced in 2009 and there are several databases hosting information on this important crop. These include Maize GDB based on GBrowse, and maizesequence based on EnsEMBL.

Rice was one of the first crop genomes to be sequenced and there are now numerous resources available to mine this genomic information. Oryzabase is an integrated rice science database established in 2000. The database hosts information on genetic resources, chromosome maps, genes and rice mutants. This is complemented by a rice genome annotation project which presents data using GBrowse.

Although wheat is an extremely important crop, advances in genomics have been limited by its large and highly complex genome. Assemblies of the gene rich regions for the group 7 chromosomes have been completed, and annotated sequences, including a large number of SNP polymorphisms are available.

A central portal for Brassica data is maintained at Brassica.info, with links to genetic marker, map and a range of diverse Brassica related information. The recently sequenced Brassica rapa genome is hosted in a database named BRAD, with a second database which contains Brassica repeat information. Both of these databases use GBrowse.

The Legume Information System supports basic research in the legumes by relating data from multiple crop and model species, and by helping researchers traverse among various data types. It currently hosts data for seventeen species and includes GBrowse databases for Glycine max (soybean), Lotus japonicus (birds foot trefoil), Medicago truncatula (barrel medic) and Cajanus cajan (pigeonpea). Lis is complemented by detailed soybean data hosted at SoyBase.

The SOL Genomics Network (SGN) is a clade oriented database containing genomic, genetic, phenotypic and taxonomic information for plant genomes, with a focus on the Euasterid clade, which includes Solanaceae (e.g., tomato, potato, eggplant, pepper and petunia) and Rubiaceae (coffee). As well as being a resource for basic crop research, SGN maintains databases with a specific focus on giving breeders direct links to breeder-relevant tools and data.

There are currently a range of databases dedicated to generic genome data or focusing on specific crops or clades. Both the type and volumes of data have increased greatly over the last few years and this trend looks to continue. Some of the early database formats are either no longer used or have limited applications, however several newer web tools are now becoming predominant. These include the GBrowse genome viewer and associated open source bioinformatics developments as well as the EnsEMBL system. As genome technology continues to advance and an increasing number of crop genomes become available, an expanding number of these databases will be developed. One of the main challenges facing crop bioinformatics researchers is to make the ever increasing volume and types of data available in a suitable format for analysis. This includes new high-throughput plant phenotype data as well as the increasing volumes of genotypic diversity data. It will be the association of this diversity data with heritable phenotypes which will likely drive genome database development over the coming years. These databases therefore will require the implementation of appropriate statistical tools for association of high-density genotype and high throughput phenotype data.

SEQUENCING, REASSEMBLY AND ANNOTATION OF ARABIDOPSIS

Arabidopsis thaliana was the first plant, and the third multicellular organism after Caenorhabditis elegans and Drosophila melanogaster, to be completely sequenced. At the time, it was claimed that the Arabidopsis genome sequence "creates the potential for direct and efficient access to a much deeper understanding of plant development and environmental responses, and permits the structure and dynamics of plant genomes to be assessed and understood." Five years on, how justified was this claim? Furthermore, a vision for the Arabidopsis research community was articulated based on the promise of the genome sequence. A noteworthy aspiration was to "determine the function of all Arabidopsis genes by 2010." What progress has been made toward this goal?

Progress in Sequencing, Reassembly and Annotation of the Genome

Since systematic sequencing was completed in late 2000, the genome sequence has undergone several rounds of reassembly, hole patching, and extension into unsequenced regions. The sequenced and analyzed regions of the genome cover ~119 Mb (million base pairs) of genome sequence. The most impressive accomplishment has been to extend BAC and YAC contigs further into pericentromeric regions. The size of the remaining gap in each chromosome, estimated using gel electrophoresis, varied between 4 Mb and 9 Mb, yielding an overall genome size of ~146 Mb. Several new genes were also identified in the pericentromeric heterochromatin. These regions are among the most comprehensively described and are a key resource for understanding and using centromere functions, for example, to make mini chromosomes, for understanding the biological functions of repeat sequences and how they originate and are maintained.

The initial set of gene models was generated by a combination of optimized ab initio gene-finding algorithms and supporting data such as EST and cDNA sequence. This was carried out in several locations (for maximum speed) and inevitably discrepancies occurred. Subsequently more systematic rounds of annotation incorporated a large number of full-length (FL) cDNA sequences such that 16,000 of the 29,000 predicted genes are supported by experimental evidence. Comparison of Arabidopsis sequences with genomic sequence from the closely related Brassica oleracea (Chinese cabbage) identified regions of high similarity that either identified putative new genes or extended existing gene models. About 30% of these new genes encoded a transcript. About 25% of the originally predicted genes had no supporting evidence such as an EST match or reasonable similarity of their putative peptide sequence to any other protein (the hypothetical genes), and consequently their biological relevance remains doubtful. Nevertheless, systematic analysis of this class of genes on Chromosome 2 revealed that the majority of these were expressed. Other analyses of EST matches to the genome sequence found several hundred matches to putative noncoding regions and other noncanonical gene-like entities.

The canonical genome sequence and annotation is currently represented by the TIGR (The Institute for Genome Research) v5 analysis. A comprehensive reassembly of the genome incorporated newly sequenced BACs from centromeric regions and carefully examined clone overlaps to yield 119 Mb of analyzed sequence. Genes in this new assembly were then predicted using FL-cDNA and EST sequences, resulting in 19,117 genes matching assemblies of EST and FL-cDNA sequences. About 800 new genes were identified in intergenic space. A total of 26,207 protein-coding genes encode 27,885 distinctive proteins by alternative splicing, nearly all of which contain known protein domains. Approximately 36% of the predicted proteome is encoded by segmental and tandem genomic duplications. Approximately 1400 pseudogenes were identified, many of which had degenerated protein sequences and premature stop codons compared to the family members to which they were most closely related. Improved analysis of repeat and transposon sequences helped to identify 2355 transposon loci. Many of these had been defined as protein-coding genes in the original annotation. Table summarizes some of the analyses of the TIGR v.5 annotation and other data.

Genome size	146 Mb (estimated)
Sequenced and annotated genome space	119 Mb
Predicted protein coding genes	26,207
Alternately spliced genes	2330
Protein coding genes with identified transcripts	19,117
Genes in protein families	18,641
Transposons and pseudogenes	3786
Distinct proteins	27,855
Nonredundant cloned ORFs	14,668
Genes with insertions in exon + intron space	24,589

The new set of predicted genes was functionally annotated using conserved domain composition, not overall sequence homology as performed originally (The Arabidopsis Genome Initiative 2000). This permits protein families and relationships between proteins to be more thoroughly defined. Gene Ontology terms were used to describe each gene in terms of the molecular function of the encoded protein, the biological process in which the protein functions, and the cellular component to which the protein may belong. These manual and computational assignments will continue to be refined as further evidence of gene functions are determined.

An Affymetrix whole-genome array (WGA) for the Arabidopsis genome was hybridized with cRNA to detect transcripts and their chromosomal location. This landmark analysis supported ~5000 hypothetical gene models as actively transcribed genes, and identified transcripts arising from many intergenic regions and from 20% of the 1300 pseudogenes. Several transcriptional hotspots were identified in centromeric regions that corresponded to a variety of repeats and transposons as well as previously unrecognized genes. Finally, transcription units for most of the genes were accurately defined and used to identify full-length ORF clones for >30% of the genome. These ORFs have been cloned (by the RIKEN/SALK and Agrikola projects) and are critically important resources for functional genomics.

The Dynamic Genome

One of the major features of the Arabidopsis genome revealed by the genome sequence was the extent of gene duplication and segmental duplications, which was surprising given the expectation of a functionally compact genome. Approximately 60% of the genome was thought to be derived from a single duplication event, possibly of the entire genome (The Arabidopsis Genome Initiative 2000). Subsequently, more detailed analysis proposed that the Arabidopsis lineage has undergone at least two duplications, the most recent being a polyploidization event during the early evolution of the crucifers (Blanc and Wolfe 2004b). These analyses support a model of Arabidopsis genome evolution involving cycles of gene duplication, gene loss, and gene divergence. Genes that remain duplicated tend to become specialized—for example, by different expression patterns (Blanc and Wolfe 2004a). The frequent observation of gene families as tandem arrays further supports the view of a dynamic genome driven by duplication events and specializing gene function from multiple copies of genes. The extensive work carried out based on the Arabidopsis genome sequence also supports interpretations of the evolution of the vertebrate lineage that propose a central role for genome duplications. Future work aims to assess and understand genome dynamics, such as gene loss, and altered gene expression that occurs as a consequence of polyploidization. These studies draw on the Arabidopsis genome sequence to understand how genome duplications shape evolution and crop plant performance.

A. thaliana is a wild species adapted to survive in a wide geographical range, and there is a long legacy of its use as a model for adaptation. Consequently there is an extensive range of natural variation in growth and environmental responsive traits that provide an exceptionally rich source of diversity. Several loci exhibiting variation in complex traits (Quantitative Trait Loci or QTL) have been cloned. Examples include using linkage disequilibrium (LD) to fine map the FRI and FLC loci controlling flowering time. Natural variation in hypocotyl responses to light were shown to be due to polymorphisms in phytochrome light receptors. Frequent sequence polymorphisms were identified between the canonical sequenced strain Columbia (Co) and the extensive sequence of another lab strain, Landsberg erecta (Ler) (The Arabidopsis Genome Initiative 2000), that are a key resource for map-based cloning. High-throughput methods for identifying polymorphisms have been developed, such as capillary-based SNP detection, to exploit the useful polymorphisms. Affymetrix expression arrays have also been used for genotyping; total genomic DNA from Recombinant Inbred Lines (RILs) made from a cross of Col and Ler was hybridized to the ATH1 Affymetrix array, and recombination events were identified. Recently, Nordborg et al. sequenced small genomic regions of 96 accessions of A. thaliana to provide a systematic survey of a dense pattern of polymorphisms from a large sample of individuals. This allowed, for the first time, a view of patterns of genome-wide polymorphism and population structure. A clear relationship between geographical distance and genotype distance was shown by clustering allele frequencies. This was quite surprising given that humans have transported Arabidopsis around the globe. There is widespread sharing of variation within these populations, and common accessions (ecotypes) share haplotypes though recombination, demonstrating that these lab strains must not be considered as "lineages." The statistical tests commonly used to analyze polymorphism data are not valid in A. thaliana because there is "too much local polymorphism." For example, there was a positive correlation between polymorphism levels and genome duplications, and a negative correlation with gene density. These factors require that studies to identify polymorphism frequencies that contribute to phenotypic variation will require surveys of the whole genome. Nevertheless, A. thaliana is a good model for evolutionary genomics. Recently, projects were announced by the DOE Joint Genome Institute (JGI) that aim to sequence the genomes of Arabidopsis lyrata and Capsella rubella, which are close relatives of A. thaliana, using a whole-genome shotgun strategy.

These sequences will enable significant advances in many aspects of A. thaliana biology, for example, by defining ancestral polymorphisms and conserved regions or "footprints" that may have functional significance.

References

- NexGenEx-Tom: a gene expression platform to investigate the functionalities of the tomato genome. BMC Plant Biol. 2015;15(1):48
- The map-based sequence of the rice genome. The map-based sequence of the rice genome. Nature 2005, 436, 793–800
- Adams, M.D., Celniker, S.E., Holt, R.A., Evans, C.A., Gocayne, J.D., Amanatides, P.G., Scherer, S.E., Li, P.W., Hoskins, R.A., Galle, R.F., et al. 2000. The genome sequence of Drosophila melanogaster. Science 287: 2185-2195
- Rice transposable elements: A survey of 73,000 sequence-tagged-connectors (STCs). Genome Res. 10, 982–990

Permissions

All chapters in this book are published with permission under the Creative Commons Attribution Share Alike License or equivalent. Every chapter published in this book has been scrutinized by our experts. Their significance has been extensively debated. The topics covered herein carry significant information for a comprehensive understanding. They may even be implemented as practical applications or may be referred to as a beginning point for further studies.

We would like to thank the editorial team for lending their expertise to make the book truly unique. They have played a crucial role in the development of this book. Without their invaluable contributions this book wouldn't have been possible. They have made vital efforts to compile up to date information on the varied aspects of this subject to make this book a valuable addition to the collection of many professionals and students.

This book was conceptualized with the vision of imparting up-to-date and integrated information in this field. To ensure the same, a matchless editorial board was set up. Every individual on the board went through rigorous rounds of assessment to prove their worth. After which they invested a large part of their time researching and compiling the most relevant data for our readers.

The editorial board has been involved in producing this book since its inception. They have spent rigorous hours researching and exploring the diverse topics which have resulted in the successful publishing of this book. They have passed on their knowledge of decades through this book. To expedite this challenging task, the publisher supported the team at every step. A small team of assistant editors was also appointed to further simplify the editing procedure and attain best results for the readers.

Apart from the editorial board, the designing team has also invested a significant amount of their time in understanding the subject and creating the most relevant covers. They scrutinized every image to scout for the most suitable representation of the subject and create an appropriate cover for the book.

The publishing team has been an ardent support to the editorial, designing and production team. Their endless efforts to recruit the best for this project, has resulted in the accomplishment of this book. They are a veteran in the field of academics and their pool of knowledge is as vast as their experience in printing. Their expertise and guidance has proved useful at every step. Their uncompromising quality standards have made this book an exceptional effort. Their encouragement from time to time has been an inspiration for everyone.

The publisher and the editorial board hope that this book will prove to be a valuable piece of knowledge for students, practitioners and scholars across the globe.

Index

www.ingramcontent.com/pod-product-compliance
Lightning Source LLC
LaVergne TN
LVHW080632210926
848485LV00003B/248